Catherine McLoughlin, PhD
Acram Taji, PhD
Editors

Teaching in the Sciences
Learner-Centered Approaches

Pre-publication
REVIEWS,
COMMENTARIES,
EVALUATIONS . . .

"There are many impressive features of *Teaching in the Sciences: Learner-Centered Approaches.* The recognition of the need for greater emphasis to be focused on tertiary science students as learners is highly commendable and a refreshing change from current practices. This book clearly recognizes that there are understandings, based on sound educational research, of how students best learn. Most important for tertiary science educators, exemplars are given showing how this information can translate into their teaching approaches.

The book explores the full scope of issues associated with tertiary science teaching and offers guidance in how to approach the teaching of content in the various science disciplines, assessment, the laboratory as a teaching/learning context, and the embracing of new technologies. This diversity and exemplification are strengths of this book.

No modern tertiary educator can afford to ignore issues associated with their teaching. For too long tertiary science educators have been appointed on the basis of their research capabilities with little or no emphasis on their teaching abilities or their understanding of the learner. This book is a wonderful resource for tertiary science educators to explore a diverse range of approaches in which the needs of the learner are the prime objective. Adoption of the strategies exemplified in this book will ensure that their teaching is based on very sound, contemporary educational principles. I have no hesitation in strongly recommending this book to tertiary science educators."

Bruce G. Cameron, PhD (UNE), DipEd, BSc (Hons)
*Senior Lecturer in Science
and Environmental Education,
University of New England,
Armidale, NSW, Australia*

Food Products Press®
An Imprint of The Haworth Press, Inc.
New York • London • Oxford

Teaching in the Sciences
Learner-Centered Approaches

FOOD PRODUCTS PRESS®
Crop Science
Amarjit S. Basra, PhD
Senior Editor

Handbook of Formulas and Software for Plant Geneticists and Breeders edited by Manjit S. Kang

Postharvest Oxidative Stress in Horticultural Crops edited by D. M. Hodges

Encyclopedic Dictionary of Plant Breeding and Related Subjects by Rolf H. G. Schlegel

Handbook of Processes and Modeling in the Soil-Plant System edited by D. K. Benbi and R. Nieder

The Lowland Maya Area: Three Millennia at the Human-Wildland Interface edited by A. Gómez-Pompa, M. F. Allen, S. Fedick, and J. J. Jiménez-Osornio

Biodiversity and Pest Management in Agroecosystems, Second Edition by Miguel A. Altieri and Clara I. Nicholls

Plant-Derived Antimycotics: Current Trends and Future Prospects edited by Mahendra Rai and Donatella Mares

Concise Encyclopedia of Temperate Tree Fruit edited by Tara Auxt Baugher and Suman Singha

Landscape Agroecology by Paul A. Wojkowski

Concise Encyclopedia of Plant Pathology by P. Vidhyasekaran

Molecular Genetics and Breeding of Forest Trees edited by Sandeep Kumar and Matthias Fladung

Testing of Genetically Modified Organisms in Foods edited by Farid E. Ahmed

Agrometeorology: Principles and Applications of Climate Studies in Agriculture by Harpal S. Mavi and Graeme J. Tupper

Concise Encyclopedia of Bioresource Technology edited by Ashok Pandey

Genetically Modified Crops: Their Development, Uses, and Risks edited by G. H. Liang and D. Z. Skinner

Plant Functional Genomics edited by Dario Leister

Immunology in Plant Health and Its Impact on Food Safety by P. Narayanasamy

Abiotic Stresses: Plant Resistance Through Breeding and Molecular Approaches edited by Muhammad Ashraf and Philip John Charles Harris

Multinational Agribusinesses edited by Ruth Rama

Crops and Environmental Change: An Introduction to Effects of Global Warming, Increasing Atmospheric CO_2 and O_3 Concentrations, and Soil Salinization on Crop Physiology and Yield by Seth G. Pritchard and Jeffrey S. Amthor

Teaching in the Sciences: Learner-Centered Approaches edited by Catherine McLoughlin and Acram Taji

Durum Wheat Breeding: Current Approaches and Future Strategies edited by Conxita Royo, M. M. Nachit, N. Di Fonzo, J. L. Araus, W. H. Pfeiffer, and G. A. Slafer

Teaching in the Sciences
Learner-Centered Approaches

Catherine McLoughlin, PhD
Acram Taji, PhD
Editors

Food Products Press®
An Imprint of The Haworth Press, Inc.
New York • London • Oxford

Published by

Food Products Press®, an imprint of The Haworth Press, Inc., 10 Alice Street, Binghamton, NY 13904-1580.

Cover design by Marylouise E. Doyle.

Library of Congress Cataloging-in-Publication Data

Teaching in the sciences : learner-centered approaches / Catherine McLoughlin, Acram Taji, editors.
    p. cm.
  Includes bibliographical references.
  ISBN 1-56022-263-8 (alk. paper)—ISBN 1-56022-264-6 (soft : alk. paper)
  1. Science—Study and teaching (Higher) 2. Active learning. I. McLoughlin, Catherine. II. Taji, Acram.
Q181.T3515 2004
507'11—dc22

2004006959

CONTENTS

ABOUT THE EDITORS

Catherine McLoughlin, PhD, is head of the School of Education at Australian Catholic University in Canberra, Australia. Her areas of specialty in teaching and research include research and development in flexible and online learning, innovative pedagogy in higher education, curriculum design, and assessment strategies.

Dr. McLoughlin is the editor of the *Australian Journal of Educational Technology* and a member of the Program Committee of the World Conference on Educational Multimedia and Hypermedia, organized by the Association for Advancement of Computing in Education (AACE). She has authored more than 100 refereed chapters and articles in international journals.

Acram Taji, PhD, is Professor of Horticultural Science at the University of New England (UNE) in Armidale, New South Wales, Australia, where she also supervises honors and postgraduate students at the MSc and PhD levels. She has authored more than 150 journal and conference papers and is the author of *In Vitro Plant Breeding* (Haworth), one of six books she has authored or edited on plant propagation and tissue culture.

Dr. Taji has twice been honored with the UNE's Vice-Chancellor's Award for Excellence in Teaching, won the inaugural Australian Award for University Teaching in 1997, and was awarded the Australian Society of Plant Physiologists prize in 1998 for her contributions to teaching, learning, and scholarship.

CONTRIBUTORS

Susan Barker holds a bachelor of science from the University of Salford, a master of science from the University of Durham, a PhD from the University of Lancaster, and a postgraduate certificate in education from Manchester University in the United Kingdom. At present she is affiliated with the Department of Secondary Education at the University of Alberta in Edmonton, Canada. Susan started out as a woodland ecologist, became interested in the human impact on woodlands and how people value and view woodlands, and then progressed into ecological education and science communication. She has written widely on her interdisciplinary research. She held a British Association Media Fellowship in 2001 and as part of that worked for BBC television on environmental news reporting. Susan is a member of the World Conservation Union (IUCN) Commission on Education and Communication and is the teaching ecology secretary of the British Ecological Society. She regularly works overseas, mainly in Eastern Europe and Latin America, and is currently embarking on a biodiversity education project in West Africa. She believes that international collaboration is hugely beneficial for all parties and that we should be aware of the environmental impact of our work and the need to live sustainable lifestyles.

Philip L. R. Bonner completed a bachelor of science in biochemistry in 1978 from the University of Sussex in England and a PhD from Nottingham Polytechnic in the area of the biochemistry of seed germination in 1984. He worked as a postdoctoral fellow at Bristol and Lancaster Universities and at Central Lancashire Polytechnic in many areas of biochemistry. In 1991 he was appointed a biochemistry lecturer at The Nottingham Trent University where he currently is a senior lecturer with research interests in transglutaminases (particularly plants). Bonner utilizes ICT in his teaching to help promote student learning and to alleviate some of the problems associated with increased student numbers in higher education.

Andrew J. Boulton is an associate professor in aquatic ecology at the University of New England (UNE) in Australia. He has taught in universities in Victoria, South Australia, France, and Arizona. Andrew has a bachelor of science (honors) from the University of Western Australia, majoring in zoology, and a PhD in aquatic ecology from Monash University. Although his main research interests are in river ecology and conservation (two books and over 100 refereed papers), he has also written articles on teaching experiences and reflection for several educational journals as well as several general papers about river and biodiversity issues for the *Australian Science Teachers' Journal.* He has a graduate diploma in higher education (UNE, 2000) and was awarded a Vice-Chancellor's Award for Excellence in Team Teaching at UNE in 1996. He enjoys teaching at all levels, from *Streamwatch* with primary school students through to supervising postgraduate student theses, and takes inordinate pleasure from his students' achievements and understanding.

Sue Franklin is a senior lecturer and director of first-year biology at the University of Sydney. She has a BSc (honors) in zoology from the University of Wales (Aberystwyth), an MSc in oceanography from the University of Southampton, United Kingdom, a PhD in marine biology from the University of Sydney, and a certificate of higher education from the University of New South Wales. Sue has taught undergraduate students at the universities of New South Wales, Western Australia, and Sydney, encompassing a wide range of biological disciplines and diverse teaching methods. In 1995, she was awarded a University of Sydney Excellence in Teaching Award and in 2000 was a member of the small first-year biology team awarded a Vice-Chancellor's Outstanding Teaching Award: First-Year Teaching. In 1998, she was awarded one of the four annual software awards from the Australasian Society for Computers in Learning in Tertiary Education.

Janet Gorst holds a bachelor of science in agriculture (honors) and a master of science from the University of New England (Australia), a PhD from the Australian National University, and a graduate certificate in higher education from Griffith University. She has been involved with plant science education for eighteen years. Janet has had a peripatetic career, having worked in university, government, and commercial institutions. She was a lecturer in plant science and plant biotechnology at the University of Tasmania (Hobart) from 1991 to 1995 and at Griffith University (Nathan Campus) from 1999 to 2002 and has also taught numerous short courses in plant tissue culture at technical and further education colleges in Canberra, Newcastle, and Brisbane. She was awarded a teaching merit certificate from the University of Tasmania in 1994 and again in 1995 and received the Australian Society of Plant Scientists' Annual Teaching Award in 2001. She is currently working in the commercial world.

Rowan W. Hollingworth holds a bachelor of science (honors) and a PhD in physical chemistry from the University of Sydney. At present he teaches in the School of Biological, Biomedical, and Molecular Sciences at the University of New England (Australia), where he works mainly with distance education students in chemistry. He enjoys the challenges of creating effective teaching materials in different media to assist these students and using ICT to improve communications with and between these students. His interest in chemical education, as opposed to chemistry, grew during a period of almost a decade working on Australian government aid projects assisting the development of science departments in a number of universities in Indonesia. His research interest now is in chemical education in general and problem solving in particular. Rowan emphasizes practical ways to help with the difficult task of learning chemical concepts and solving chemistry problems. He was awarded the University of New England's Vice Chancellor's Award for Teaching Excellence in 2000.

Susan Lee completed a BA in biological sciences at Macquarie University, a research master's degree at Griffith University, and a PhD at the University of Queensland. After working as a scientific officer in government analytical laboratories, in hospital pathology laboratories, and in tertiary institutions, she became a lecturer at Griffith University in 1992, with special responsibilities for laboratory classes that supplement and illustrate the theoretical material presented to the students. Susan is particularly interested in helping students understand practical applications, improving problem-solving skills by integrating and applying knowledge gleaned from several disciplines, and developing skill and confidence in performing practical scientific investigations. Susan has been nominated for the Griffith University Excellence in Teaching awards every year since the award's inception and received certificates of commendation for her teaching skills.

Lisa Lobry de Bruyn holds a bachelor of science (with first-class honors) and PhD from The University of Western Australia and a Certificate in Higher Education from The University of New England (Australia). Lisa is a senior lecturer and has been involved in teaching many areas of natural resources at tertiary level in The University of New England since 1993. Her innovative teaching methods have been recognized and showcased in four Australian University Teaching Council grants over the last five years. She is the author of seventy-five journal and conference papers encompassing a wide range of interests including soil agroecology, soil condition monitoring, ethnopedology, natural resource management, and teaching and learning practice, as well as being invited to contribute to several special journal editions on soil agroecology, ethnopedology, and sustainability indicators. Lisa is committed to engaging with students in the process of learning to enable them to realize their full potential and to approach life's challenges with confidence and inquisitiveness.

Robyn Muldoon is the academic skills coordinator at the University of New England (UNE) in Australia. She has been responsible for developing, implementing, and promoting a range of learning support programs and resources for students at UNE. In 2001 her team won the Vice-Chancellor's Teaching Excellence Award at UNE. The tUNEup University Preparation Program, which she spearheaded, won an Australian Award for University Teaching in 2002 in the institutional category for an innovative and practical approach to the provision of student learning support services. In 2003 she was awarded the Vice-Chancellor's Award for Excellence in Equity for implementing strategies at UNE that have increased access to tertiary education for many students and improved opportunities for successful tertiary study for many more. Her research interests include discipline-based academic and professional skills programs, graduate attributes, and student mentoring programs.

Debra L. Panizzon is a senior lecturer in science education at the University of New England (UNE), Australia. Her qualifications include a diploma in teaching, bachelor of education (honors), a PhD in science education, and a graduate certificate in science education.

She received a University Medal for her honors thesis and an Award for Excellence in Research from the New South Wales Institute for Educational Research (IER) for her doctoral thesis. Currently she is teaching environmental science to bachelor's of education (primary) students and science curriculum units to preservice secondary science teachers. Prior to coming to UNE, Debra taught science, senior biology, and physics in secondary schools in New South Wales and South Australia. Her research interests are in the areas of cognition and concept acquisition in secondary and tertiary students. She also works with developmental-based assessment and the ways in which primary, secondary, and tertiary teachers use assessment to improve instruction. She has published in international and national science education journals, including the *Australian Science Teachers' Journal.*

Mary Peat holds a BSc (honors) in zoology and comparative animal physiology from the University of Birmingham, United Kingdom, and a PhD from the University of Bristol. Mary has been teaching at the tertiary level for more than thirty years, primarily at the first-year level, working with students in transition. Currently she holds the position of associate professor of biological sciences at the University of Sydney. Mary's teaching and research have been recognized with a number of national and international awards for excellence, including two University of Sydney awards and three software awards. In addition, Mary is an associate dean in the faculty of science and has been instrumental in leading changes in teaching and learning during the past decade. As the current director of UniServe Science <http://science.uniserve.edu.au/> Mary has also helped restructure some aspects of teaching science at the secondary level in New South Wales.

Frances Quinn holds a bachelor of science (honors) from the Australian National University and a graduate diploma of education from the University of Canberra. Frances is a lecturer in the Academic Skills Office at the University of New England in Australia, where she is involved in supporting the learning needs of on- and off-campus students. She has also taught in the faculty of the sciences for many years, with particular involvement in first-year biological sciences. Her current research focus is on aspects of learning in tertiary science. At present she is busy writing her doctoral thesis on learning approaches and learning outcomes in a first-year biology context.

Robyn Smyth holds a bachelor of arts and a diploma of education from Macquarie University in Australia. Her master of education degree, which included research into curriculum design, is from the University of Southern Queensland in Australia. Robyn's research for the degree of doctor of philosophy, which was awarded from the University of New England in Australia, is in the area of management of educational change. At present she works as an educational designer and staff developer in the Teaching and Learning Centre at the University of New England (UNE) in Australia. Her research interests include managing educational change, approaches to academic staff development, curriculum mapping and design, and the pedagogy of distance and face-to-face learning. She is currently engaged in researching appropriate pedagogical approaches to the use of broadband Internet-based videoconferencing. In recognition of her work implementing new curricula in secondary schools, Robyn was awarded a New South Wales Department of Education and Training Minister's Award for Excellence in Teaching and a National Excellence in Teaching Award.

Charlotte Taylor is a lecturer in the School of Biological Sciences at the University of Sydney. She holds an honors degree in botany from Dundee University, Scotland, a PhD in tropical botany from Aberdeen University, Scotland, and a master's in higher education from the University of New South Wales in Australia. She has taught, administered, and developed curricula for a range of courses in first-year biology at the University of Sydney, generally with enrollments of 900 to 1,600 students. Charlotte's involvement in curriculum design and teaching development has resulted in the receipt of an excellence in teaching award. She maintains a connection with tropical biology through research projects on pollination. In addition, she is involved in collaborations with biologists and educational developers in research projects on student learning, particularly in the areas of academic literacy, scientific writing, and communication in online learning environments. Education projects have been published and presented at educational conferences in Australia and the United Kingdom.

Foreword

As noted in the Introduction by Catherine McLoughlin and Acram Taji, something of a revolution has been building in pedagogy—the art and science of teaching—since the 1980s. Greater concentration on researching both teaching and learning has seen new and more complex understandings emerge and combine to produce firmer consensus. Terms such as *authentic pedagogy* (Newmann and Associates, 1996), *productive pedagogies* (The State of Queensland Department of Education, 2001), and *quality teaching* (Dinham, 2002) have been labels attached to such syntheses. In all of this, the role of the individual teacher has increasingly been found to be crucial in determining student achievement (see Darling-Hammond, 2000).

We now have a clearer understanding of what effective teachers do and how students learn. However, to a large extent, these understandings remain an idealized form when we consider what actually happens in many educational settings.

In a recent and comprehensive examination of the teaching of science in Australian schools, it was found that despite these new understandings and related efforts to transform curricula and teaching practice,

> the actual curriculum implemented in most schools is different from the intended curriculum. In some primary schools, often science is not taught at all. When it is taught on a regular basis, it is generally student-centred and activity-based, resulting in a high level of student satisfaction. When students move to high school, many experience disappointment, because the science they are taught is neither relevant nor engaging and does not connect with their interests and experiences. Traditional chalk-and-talk teaching, copying notes, and "cookbook" practical lessons offer little challenge or excitement to students who take science subjects in the post-compulsory years of schooling. (Goodrum, Hackling, and Rennie, 2001, p. viii)

There are well-documented pressures on teachers deriving from constant change, imposed administrative responsibilities, curricula overloaded with "content," standardized testing, external examinations, increased social expectations on schooling, lack of time and resources for professional development, and a general decline in teacher status (Dinham and Scott, 2000). This context militates against innovation in teaching and creates in students a conception of learning that is narrowly based on information transmission, reception, processing, repetition, and regurgitation. Many teachers do manage to overcome these structures to create rich and engaging learning environments, but they tend to be swimming against the tide (Ayres, Dinham, and Sawyer, 2000).

University teachers are under similar and often even greater pressures, with contracting resources and a reduced and aging staff profile. Introducing pedagogic change into this context also is difficult. Students come from high schools with traditional orientations to learning and have equally traditional expectations of the teaching they will encounter in university courses. Such attitudes can foster dependence and hinder the acquisition of deep and lasting knowledge and understanding. To compound the issue, many university teaching staff lack formal qualifications in teaching and teach as they were taught, adopting teacher-centric instructional approaches, particularly with large, first-year undergraduate groups. Teaching may not be sufficiently valued as an activity by either these teachers or university hierarchies, with research, grant acquisition, publication, doctoral completion, and tenure seen as central and all-important. The contributors to this book provide many examples of the dysfunctional consequences of this context, including lack of student engagement and high failure and dropout rates. The imperative for change is clear.

Moving from traditional modes of instruction to more student-centered approaches is not meant to downplay the role or importance of teachers. Some have simplistically and erroneously noted how the "sage on the stage" needs to be replaced with the "guide by the side." Student-centered approaches to teaching actually require greater teacher expertise, professionalism, and input. In a study of highly successful senior secondary teachers, we adopted "teacher-directed, yet student-centered" as the most apt label for what we found in the classrooms of highly accomplished teachers (Ayres, Dinham, and Sawyer, 2000).

Teachers, and high-quality teachers at that, are needed more than ever. For example, rather than making teachers redundant as some have suggested, the World Wide Web has made teachers even more central and important if students are to navigate, access, and effectively utilize the material they need for their learning. "Dumping" teaching materials on the Web is both a poor and expensive substitute for quality teaching. As Goodrum, Hackling, and Rennie (2001, p. ix) concluded from their study of science teaching, "teachers are the key to change."

How is this change in thinking and practice to be achieved, and how might it look? Some of the answers to these questions lie in this collection of thoughtful, frank, and empirically based chapters. The accounts provided are a real and valuable attempt to close the gap between cutting-edge thinking and research on teaching and learning and what still happens in many educational settings.

What is described is neither idealistic nor perfect, as the various authors are at pains to point out. Neither is it entirely new, as others would like to claim for themselves. What is described here has its seeds in the timeless aspects of good teaching—it is just that some of the contexts are new and challenging. In many cases, these accounts represent the first steps on an uncertain journey to transform and enrich teaching and learning, but the progress to date is encouraging. Here are real educators engaged with real students in the pursuit of powerful, effective, and lasting learning.

Overall, this book makes a valuable contribution to furthering our knowledge of teaching and learning. Although largely centered in the context of undergraduate science teaching, there are valuable lessons here for educators from all areas who share an interest in fostering intellectual quality and capacity in their students. The co-editors and contributors are to be commended for their efforts, both to transform their own teaching and to share their experiences with others.

Stephen Dinham
Chair of Teacher Education, Pedagogy,
and Professional Development, School of Education,
University of New England, Australia

REFERENCES

Ayres, P., Dinham, S., and Sawyer, W. (2000). Successful senior secondary teaching. *Quality Teaching Series,* No. 1, Australian College of Education, September, pp. 1-20.

Darling-Hammond, L. (2000). Teacher quality and student achievement: A review of state policy evidence. *Education Policy Analysis Archives,* 8(1): 1-22.

Dinham, S. (2002). NSW quality teaching awards—Research, rigour, and transparency. *Unicorn,* 28(1): 5-9.

Dinham, S. and Scott, C. (2000). Moving into the third, outer domain of teacher satisfaction. *Journal of Educational Administration,* 38(4): 379-396.

Goodrum, D., Hackling, M., and Rennie, L. (2001). *The status and quality of teaching and learning of science in Australian schools.* Canberra, Australia: Department of Education, Training, and Youth Affairs.

Newmann, F. and Associates (1996). *Authentic achievement: Restructuring schools for intellectual quality.* San Francisco: Jossey-Bass.

The State of Queensland Department of Education (2001). *The Queensland School Reform Longitudinal Study final report (QSRLS).* Brisbane, Australia: Education Queensland.

Introduction

Learner-Centered Approaches in the Sciences

Catherine McLoughlin
Acram Taji

Over the past two decades, advances in cognitive and pedagogical research have made great strides in helping us to understand the learning process. Findings from studies of student learning have provided insights and directions for improving educational practice. The implications of this research enable us to adopt well-established theoretical principles and theories in teaching students and adopting learner-centered principles for instruction. White (2002) describes the changes that have taken place in the sciences as a "revolution." Important features of the revolution in science teaching and research have been a growing concern for practice, the social dynamics of learning, and the complexities of students' conceptions and misconceptions of scientific concepts. New lines of research that emerged in the 1980s and 1990s included investigations of the metacognitive skills of students, cognitive preferences and learning styles, situated cognition applied to pedagogy, and the application of information and communications technologies for the creation of improved learning environments in science (e.g., Baird, 1986; Hewson and Hewson, 1983).

Within the broad topic of research on teaching and learning, emphasis has shifted during the past twenty years. Although Piagetian psychology flourished during the 1970s and early 1980s in research on science instruction, more recent studies have taken a broader view of influences and factors that impinge on learning and focus more on learning environments and social interaction patterns, assessment processes, and perceptions of teachers and students.

As the research agenda on teaching in the sciences is rich and complex, it forms the main focus of this book. In addition, we have recognized that teaching is an interventionist activity, and that there is now a well-established body of research on constructivist principles applied to teaching. Therefore, this collection of chapters from practicing teachers in higher education also has a pragmatic focus: to disseminate examples of effective pedagogy for science educators in higher education. Each of the contributors in this book has approached the central concern in a different way, and each brings different insights and perspectives to the reader. The combined chapters provide ample evidence that effective teaching in the sciences must begin with recognizing and meeting student needs, and providing environments that encourage active, strategic learners.

Many of the topics and concerns described in these chapters have been articulated in the specific context of science teaching at the tertiary level, so the book provides a unique and in-depth analysis of the issues and challenges facing science educators.

Susan Barker, in Chapter 1, "Student-Centered Ecology: Authentic Contexts and Sustainable Science," distills the major theoretical concerns of science educators in the argument for improved assessment practices, authentic learning activities, and cross-disciplinary experiences. Added to the need for more supportive teaching practices, Barker advocates evaluation and benchmarking of current practices so that constant improvement becomes a hallmark of good practice.

Mary Peat, Sue Franklin, and Charlotte Taylor, in Chapter 9, address the issue of providing constructive feedback to support learning with "Application of ICT to Provide Feedback to Support Learning in First-Year Science." The authors begin by acknowledging contemporary challenges in higher education, including large- class teaching, a diverse student body, and time pressure, that mitigate against students acting on the feedback provided to them. The authors outline an array of tried and tested strategies that enables feedback to students, including virtual learning environments, self- and peer-assessment, and online quizzes. The authors document each strategy carefully, based on more than ten years of university teaching experience in biology.

One of the most enduring issues to emerge from the research on science teaching has been a movement away from instructional approaches that emphasize mastery of content and toward constructivist pedagogies that place self-regulation of learning and strategy use as the most important ingredients in successful science learning. Studies of learning environments in science have shown that metacognition—the ability to think about thinking, to be consciously aware of oneself as a thinker, and to monitor and control one's mental processing—is essential for efficient learning. The research by Hollingworth and McLoughlin in Chapter 4 describes the development and evaluation of an online resource, metAHEAD, that scaffolds strategy development and self-monitoring in problem solving. Their chapter, "Developing the Metacognitive and Problem-Solving Skills of Science Students in Higher Education," provides a theoretical framework for metacognitive skill development and hierarchy of problem-solving tasks ranging from simple to complex and open-ended. The authors provide convincing evidence that students need these skills and cognitive strategies to enable them to solve complex problems. This can be achieved through the use of an online resource that supports reflection, process-based learning, and social dialogue in a Web forum where students participate in an online community that values constructing and communicating knowledge.

Another major area of change in the teaching of science at the tertiary level has been the adoption of a range of innovative pedagogies to support learning in the sciences. Lisa Lobry de Bruyn's chapter describes the development of a problem-based learning (PBL) approach for students in environmental science and natural resource management. Graduates are now expected to have a full range of skills and competencies along with a knowledge base and an understanding of how to work cooperatively, solve problems, and resolve conflicts. The form of problem-based learning depicted by Lobry de Bruyn in Chapter 5 is based on distributed learning and utilizes asynchronous communication tools and Web-based learning to engage off-campus students. The virtues of PBL pedagogy are made clear in this chapter, as are the design challenges faced by teachers, which include developing real-world problems and scenarios, encouraging more self-directed learning, and changing assessment practices so that students are creating knowledge rather than reproducing it. Central to this chapter are concerns about how to adopt appropriate pedagogies, how to ensure timely and relevant feedback, and how to motivate students. The chapter provides a detailed account of how a problem-based learning approach can be used to develop self-regulated learning processes among science students.

In Robyn Smyth's Chapter 11, "Exploring the Usefulness of Broadband Videoconferencing for Student-Centered Distance Learning in Tertiary Science," investigates a range of issues that address the changes needed in pedagogy and learning environments in order to ensure that learner needs are met. Smyth argues for a shift in conceptions of teaching toward "a learning paradigm" in which both teacher roles and student roles change. Essentially, for learner-centered pedagogies to really work there must be a change from teaching as knowledge transmission and didacticism to teaching as facilitating learning and self-directed inquiry. For many teachers this presents a challenge to their fundamental beliefs and professional values. The change can be effected by teachers through active research and inquiry into their own practices, with a mentor or staff developer offering feedback when appropriate. Students may also need to change their conceptions of learning and become more alert to the responsibilities they have in the learning-teaching transaction. For all stakeholders, Smyth maintains that a culture of improvement, self-renewal, and questioning are important ingredients of good teaching. This chapter also includes an interesting case study of how technology has supported a change from traditional pedagogy to learner-driven activity. Internet-based videoconferencing has been used effectively to augment and enrich communication, support group work, and personalize learning for students. The learning experiences and activities that are possible with this technology are many and varied, provided that pedagogy and student learning needs are the first priority. Smyth also provides a very useful framework in her chapter that illustrates how technology choices can support learner interactivity and constructivist pedagogy.

Robyn Muldoon's Chapter 7, "Student-Centered Learning Support in the Sciences," deals with the need to support students at all stages of study, but particularly in the first year. The chapter emphasizes that it is no longer possible to assume students are ready for "academic" study, as the diversity of the population continues to expand with the lifelong learning agenda. Muldoon provides an overview of current concerns about the learning needs of tertiary science students and approaches to supporting learning while assisting students to meet the challenges of first-year study, particularly those challenges related to tertiary literacy and learning strategies. Increasing numbers of first-year students are challenged by the expectations of tertiary institutions with respect to critical thinking, and many students struggle with their role as independent learners. If we are to improve student learning outcomes and retention rates in the sciences, it is clear that a support network is needed which utilizes the known benefits of context-specific support while also aiming to achieve academic and social integration. Muldoon's chapter provides an overview of current approaches to learning support and describes one successful approach in the Faculty of the Sciences at the University of New England. The mentoring program described is underpinned by recent research on student retention while exemplifying best practice in student learning support. It combines tried and tested features of discipline-specific academic skills development and learning support with an appropriate balance of one-to-one counseling and group work. Mentoring is a student-centered approach that runs parallel with, complements, and counterbalances traditional teaching methods to the benefit of all stakeholders, particularly students. This chapter provides a comprehensive overview of current successful approaches to learning support in the sciences and interdisciplinary teaching contexts.

Because assessment is central to the processes of learning and drives students' motivation to learn, Chapter 10, Frances Quinn's "Assessing for Learning in the Crucial First Year of University Study in the Sciences," makes an important contribution to the book. The aim of this chapter is to investigate the place of assessment within the current pedagogical context of first-year science. Some general dimensions of assessment in tertiary science education are discussed, followed by analysis of the relationship between assessment and curriculum issues, and the influence of assessment practices on the approaches students take to their learning. The chapter continues with principles for assessment of first-year students, along with examples drawn from current published case studies of assessment alternatives used in a variety of large introductory science classes. Awareness of the crucial importance of the first year at university is growing in Australia and beyond. In their transitional first year, students are forming views concerning what university learning is about that will influence their learning strategies and expectations in subsequent years. First year is also a period of high risk for student failure or attrition. The adoption of appropriate assessment strategies to support and contribute to student learning and retention in the first year is therefore crucial, as assessment is the most powerful lever teachers have to influence the way students respond to courses and behave as learners. Yet despite the pressing need for assessment that is centered on the needs of this diverse group of learners, first-year science classes traditionally have been marked by assessment strategies least supportive of student learning and motivation, such as heavy reliance on end-of-semester objective exams. The chapter aims to provide a convincing rationale of the need for innovation in assessing first-year science students and a source of ideas and further information for teachers of first-year tertiary science.

In Debra Panizzon's and Andrew Boulton's Chapter 8, titled " 'Drowning by Numbers': The Effectiveness of Learner-Centered Approaches to Teaching Biostatistics in the Environmental Life Sciences," the authors note that over the past few decades, research in teaching and learning in science has shown that students enter formal education with well-developed ideas about scientific concepts that have been formulated intuitively or through their sensory experiences with their natural world. Although these conceptions are coherent and make sense to the student, they may differ significantly from formal scientific explanations. To enable students to bridge the gap between their intuitive conceptions and the contemporary scientific view, a significant change in conceptual knowledge must take place. In the process of change, the role of the student changes from that of a *passive recipient* to an *active constructor* of knowledge.

In this chapter, Panizzon and Boulton review current educational theories about the way students learn, with specific applications in the life sciences and biostatistics. Commencing with an exploration of constructivism, they highlight the importance of social engagement and educational context. Following this is an exploration of the issue of alternative conceptions of scientific concepts in order to explain why students may retain incorrect views about scientific concepts even after formal instruction. They also outline a number of teaching strategies and learning opportunities provided to help students in survey design and biostatistics construct a deeper understanding of statistics.

Philip L. R. Bonner, in Chapter 2, addresses how information and communications technology can assist teachers in developing lifelong learners and meeting the diverse learning needs of their students. While emphasizing that technology alone cannot promote student learning, Bonner discusses the ample evidence emerging that teacher invention and judicious use of technology can foster reflection, problem solving, and independent inquiry. This chapter provides examples of a range of computer-assisted learning packages currently available and designs that encourage improved learning. Another important message in this chapter is the need to constantly evaluate teaching and learning outcomes in technology-supported environments and to use the results to improve teaching.

In Chapter 3, Janet Gorst and Susan Lee seek to examine aspects of student-centered learning in the undergraduate life sciences laboratory. They present an in-depth study of third-year biology students at an Australian university to illustrate that there is diversity in what students bring to laboratory work, what they perceive to be the reasons for and relevance of laboratory work, how they learn, and what learning difficulties they have. Teachers need to be aware of such diverse needs and perspectives and plan learning activities accordingly. Gorst and Lee emphasize the need to foster deep learning approaches and argue that laboratory course designers can encourage deeper learning by including more experiments in which the students are given some independence in designing protocols and by encouraging students to be critical, to raise their own questions, and to offer alternative perspectives.

In McLoughlin and Hollingsworth's chapter "Problem Solving in the Sciences: Sharing Expertise with Students," the authors describe how the teaching of problem solving requires the adoption of process-based pedagogies that reveal to students the ways in which experts solve problems and the coaching of students in higher-order skills that lead them away from a preoccupation with merely finding solutions toward building up a repertoire of problem-solving strategies. Chapter 6 also suggests that science educators need to model problem solving explicitly by thinking aloud and demonstrating the skills they seek to develop in their students rather than simply expecting students to become effective problem solvers without any form of specific scaffolding. The chapter provides examples and cases that illustrate such process-based approaches to student learning using examples from undergraduate chemistry programs. McLoughlin and Hollingworth present a convincing research-based argument that, through the adoption of pedagogical activities such as thinking aloud, coaching, and articulation of problem-solving processes, teachers can assist students to develop a repertoire of skills and self-awareness of their own problem-solving strategies.

REFERENCES

Baird, J.R. (1986). Improving learning through enhanced metacognition: A classroom study. *European Journal of Science and Education,* 8: 263-282.

Hewson, M.G. and Hewson, P.W. (1983). Effect of instruction using students' prior knowledge and conceptual change strategies on science learning. *Journal of Research in Science Teaching,* 20: 731-743.

White, R. (2002). The revolution in research on science teaching. In V. Richardson (Ed.), *Handbook of research on teaching* (pp. 457-471). Washington, DC: American Educational Research Association.

Chapter 1

Student-Centered Ecology: Authentic Contexts and Sustainable Science

Susan Barker

INTRODUCTION

Ecology is a subject that has given rise to more controversy regarding its place in educational systems than any other branch of science. It is a relatively new science, and in its early years it was described not as a science at all but merely a point of view (McIntosh, 1985). The environmental crisis highlighted by Rachel Carson in the 1960s (Carson, 1962) thrust ecology into the public arena, helping raise its profile and develop it into a fully fashioned science with a rich array of principles, concepts, observations, experiments, and models. Some critics of ecology, for example, di Castri and Hadley (1986), still question the validity of its status as a science and highlight the lack of rigor, weak predictive capability, and failure to harness modern technology as shortcomings. Despite these criticisms, ecology, which is very much the study of interrelationships within our environment, clearly adheres to the principles and definitions of science (Wali, 1999). Ecology was, and is, a science that does not readily fit into the familiar mold of science erected on the model of classical physics, as it deals with phenomena touching on human sensibilities, including ethics, morality, and economics. So this complex history, as well as its fuzzy boundaries, raises issues about how ecology is taught, and these challenges have long been recognized (Lambert, 1966). The goal of this chapter is to present a case for student-centered learning using authentic and contemporary approaches in which the quality of the learning experience is improved and at the same time is commensurate with the principles of sustainable development.

CHALLENGES IN TEACHING ECOLOGY

One problem facing ecology teaching is that a traditional scientific method approach to investigations/practicals and accumulation of ecological data and knowledge needs to be accompanied by ways of allowing students to make value judgments. The provision of authentic contexts for these strategies as well as demonstrating a commitment to environmentally sound and sustainable practice are essential. Authentic contexts allow students firsthand opportunities to study organisms in their natural habitats, e.g., sampling tree invertebrates in woodland. This not only provides them with original data but can also provide data that are meaningful and relevant. Moreover, the notion of environmental responsibility across all subject areas within higher education is well recognized (Toyne, 1993). The communication of this cannot be achieved using traditional transmission modes of teaching, where information pathways are unidirectional. It is also difficult to teach in the classroom or laboratory unless students are able to mentally draw on experiences and examples that they have encountered firsthand. Indeed, a number of authors worldwide highlight the cognitive, affective, and behavioral advantages of using authentic contexts for ecology (Cobern, 1993; Manzanal, Barreiro, and Jimenez 1999; Orion and Hofstein, 1994).

Ecology is the study of living organisms in their natural environment, and this encompasses the whole of the physical environment as well as the organisms themselves. Thus it would seem that the most appropriate subjects to be studied in preparation for university lie well outside a traditional science education. Although it is often a requirement to have studied sciences for entry into ecology undergraduate programs, this is an inadequate preparation for a broad discipline such as ecology, for which history and geography would provide equally valuable insights. Biology alone is often no longer an adequate preparation, as much of the traditional systematic botany and zoology has gone at the expense of more contemporary fields such as genomics. Moreover, students now entering into higher education are more varied in terms of their school experiences due to recent changes in the advanced-level school curriculum. This poses some challenges, for example, teaching a year-one introductory course in ecology when some students in the class have studied an advanced-level module in ecology at school and others not. The challenges of coping with diverse student experiences as well as increased student numbers also provides a real opportunity to improve the learning experience by putting the learners at the heart of the teaching process, where their experiences can be shared and a community of learners catered to. Rising numbers of students, diversity of student experiences, the interdisciplinary nature of ecology, and the need for authentic contexts to address knowledge skills and values all provide a clear pedagogical rationale for student-centered approaches in ecology. This chapter explores ways in which pedagogical change toward student-centered learning can improve the quality of the learning experience for all.

EXPERIENTIAL AND CONSTRUCTIVIST APPROACHES

Ecology provides us with an excellent context for interdisciplinary study, issues-based teaching, and a science education that is relevant. Consideration of the pedagogy is just as critical as decisions of content, and it should certainly be given priority over available technology. Experiential learning and constructivist approaches to teaching and learning, which are at the heart of student-centered approaches, are important strategies to help students make value judgments and to reflect on their own behavior and learning (Fensham, Gunstone, and White, 1994; Kolb, 1984; Kraft and Kielsmeier, 1995). At the core of constructivism is a "view of human knowledge as a process of personal cognitive construction, or invention, undertaken by the individual who is trying, for whatever purpose, to make sense of her social or natural environment" (Taylor, 1993, p. 268). Therefore, when teaching students about the natural environment with a view to promoting sustainable practices, constructivist teaching makes good sense. The learning process is based on the personal experiences of the students and the acquisition of knowledge is the product of activities that take place in particular cultural contexts. Knowledge is constructed by the learners in the sense that they relate new elements of knowledge to already existing cognitive structures (Bruner, 1993). Thus this approach to teaching can help overcome some of the challenges in higher education today, notably students with a wide variety of experiences, prior knowledge, and goals. By comparing the traditional science curriculum with that of constructivist approaches the benefits are further outlined (see Table 1.1).

TABLE 1.1. Comparison of Traditional and Constructivist Curriculum

Traditional science curriculum	Constructivist science curriculum
Scientific knowledge	Knowledge about science
What we know	How and why we know
Emphasizes fully developed final-form explanations	Emphasizes knowledge, growth, and explanation development
Breadth of knowledge	Depth of knowledge
Basic scientific knowledge	Conceptualized science knowledge
Curriculum units discrete	Curriculum connected

Source: After Duschl and Gitomer, 1991.

So how can increasingly large numbers of diverse students be taught through a staged constructivist approach in a way that sets good examples of sustainable development by promoting good practice? One example is a second-year ecology and conservation optional course (elective) within the Science Faculty at the University of Warwick, United Kingdom. Students taking it come from a variety of departments, e.g., mathematics, computer science, biology, chemistry, physics, statistics, and engineering. The challenges of teaching this optional course are

- diversity of prior experience and knowledge in the students;
- timetable issues, e.g., 9:00 a.m. lectures;
- physical location within the university, e.g., distance from home department;
- physical room layout, e.g., large lecture theater;
- large numbers of students (fifty to eighty); and
- resource availability for large numbers of students in a discipline that is rapidly changing and needs updating every year.

The students thus have widely varying levels of scientific knowledge, information technology (IT) skills, and expertise, while the course objectives aim to enhance all participants' understanding of ecology and sustainable development. In traditional lecture courses the communication is often presentational, the outcomes predetermined, and the learner passive, and over a number of years it became clear that the objectives were not always being achieved for all students.

Virtual Learning Environment

The course was clearly not as effective as it could have been and thus was radically reexamined, taking the pedagogical issues as the most important priorities. The introduction of a virtual learning environment (VLE) was the most significant change, with some of the traditional lectures replaced with online conferences in this forum. Here the students were required to visit the VLE and read the instructors' introduction, which set out the aims, objectives, and expectations for the session. They then had to process the lecture notes (written in a student-centered style) and complete a number of short tasks for active participation, e.g., click on a hypertext link to a specific Web site. Finally, the students consolidated and reflected on their work by contributing their views or factual material (depending on the task) to the VLE for all participants to read. This strategy helped students to become active learners (they have considerable autonomy and the mode of communication is a multifaceted dialogue). The other benefits of the text-based computer conference were of the "added-value" nature:

- It had "real-world" currency as a tool for use in commerce, industry, and the professions.
- There were increasing numbers of documented case studies of its use in professional development (Salmon, 2000).
- It was an example of sustainable practice. It cut down on traffic at peak times. It reduced the number of students' journeys during the day and handouts were printed only when needed by the student, as all of the lecture notes were available online.

The way in which students engaged in the virtual learning environment certainly has parallels with constructivist teaching approaches through a staged approach. The idea of a staged approach was formalized by Salmon following research and evaluation based on the Open University Business School's distance-education programs (Salmon, 2000). Salmon has set out a five-stage developmental model:

Stage 1: Access and motivation
Stage 2: Online socialization
Stage 3: Information exchange

Stage 4: Knowledge construction
Stage 5: Development

This very much parallels the process of constructivism as outlined earlier, in which students were motivated to contribute online. They were able to socialize online, present their own ideas, and read the ideas of other students, staff, and those presented on designated Web sites. These steps allowed them to develop their own understanding of the issues in question. This is critically important in ecology, where it is recognized that students rarely think about nature purely from a scientific perspective but from a worldview, with religion, aesthetics, conservation, and scientific contexts as common reference points (Cobern, 2000).

There is no doubt that large lecture theaters are an impediment to constructivist teaching. The links between constructivist teaching and educational technology are well documented, with a consensus that individuals engaged in learning should have the opportunity to inquire and to develop understanding from their own and others' perspectives, and that educational technology can facilitate this (Adams, 1989).

In the ecology and conservation course highlighted previously, online conferencing was used to

- reflect on lecture sessions by posing questions immediately after a lecture;
- substitute lectures with online lectures and Web-based tasks; and
- support students leading up to the exams with responses to individual queries for all to see.

Computer conferencing does not always engage the totality of any particular group, but this is also true of traditional lectures and seminars. Analysis of student engagement with the VLE reveals that about one-third of the students participate fully as contributors or as spectators or listening in; about one-third will engage to a degree, but usually as part-time spectators; and one-third will have negligible involvement, possibly never even logging on and using the VLE. Although this is rather disappointing when so much time and effort has gone into planning the conferences, it is probably no worse than engagement in seminars and definitely lectures. To encourage participation in computer conferences, online activity needs to be purposeful and embedded in the program. Achieving this mix is not straightforward, as many commentators make clear (Salmon, 2000; McConnell, 2000).

Changing Assessment Practices

An important area in which computer conferencing needs to be embedded is assessment. Students ready and willing to participate in the VLE deserve, and may need, recognition; those who are less forthcoming may be influenced by a direct connection with assessment arrangements. Students are very much influenced by the tangible outcomes of courses and how it contributes to their final degree, and often are not able to see the intangible benefits—particularly an improved understanding or other long-term benefits (Evans and Abbott, 1998). Where assessment is purely by examination, the immediate rewards to the student are not clear. The examination in this case thus was changed slightly to include one optional question related to the use of the VLE within an ecology context. The immediate effect of this was an increased participation by students in subsequent years.

Over a seven-year period of using VLEs in the ecology and conservation elective course there was evidence of some improvement in the formal assessment. Exam answers demonstrated a wider range of ecological examples but with only a marginally better mean mark in exam performance. However, from course evaluations, the longer-term benefits in terms of student attitudes toward ecology and sustainable development are likely to be significant. The course evaluations have been much more positive, with students welcoming the availability of lecture notes and more flexibility in terms of when and how they can do their learning. For example, students indicated that they do not like nine a.m. lectures and preferred an online task at a time convenient to them, e.g., "I simply cannot get into the university for nine a.m., so it is great not to feel guilty and not to miss out on notes." However, most students were clear that they would not like to see the entire course delivered through the VLE. They do like face-to-face contact with an instructor and their peers, and enjoy the performance of a lecture, e.g.,

I see real benefits of the VLE for us as students and really liked doing all the tasks as it did not seem like work. But I would not like the whole course delivered this way as I enjoyed the lectures immensely, and in fact the VLE helped me get more out of the lectures as I could read lecture notes beforehand.

Very few recognized the tremendous additional environmental benefits that using VLEs offer. In the early days of using VLEs (in 1997) there were some problems with students lacking sufficient IT skills to get the most out of the facility. This has steadily improved over the years and in 2004-2005 all students are completely comfortable with the IT demands on the VLE. Indeed, many students find this a motivating factor. The main problem now is the infrastructure within the university and the provision and access to computers.

In conclusion, the use of VLEs, particularly through a text-based computer conference and online resources, can significantly improve the quality of teaching and active learning in interdisciplinary bioscience courses *and* promote sustainable development through examples of good practice.

LEARNER-CENTERED TEACHING MATERIALS: PROMOTING SUSTAINABILITY

The use of virtual learning environments, with their potential for creating a paperless learning environment and reducing traffic congestion at peak times, means that it is a very sustainable teaching strategy. However, students still lack the experience and confidence to work solely through this medium, and many students indicate that a lecture "performance" with some paper handouts is their preferred style of learning. Learner-centered approaches are heavily reliant on teaching materials that guide the learners and act in place of, or in addition to, the instructor. Such teaching materials can be substantial, as they need to be interactive and attractive to engage the learner and thus rely heavily on the use of white space, diagrams/images, and so on (Race, 1989, 1998). In a discipline such as ecology, which professes to help protect and conserve the environment, the abundance of paper can appear hypocritical and irresponsible. Where paper materials are used, then, the manner of their production should be viewed as a learning opportunity. For example, indicating how the teaching materials have been developed in a sustainable fashion is an important teaching point. Indeed, a voluntary code established by the U.K. government panel on sustainable development aims to raise standards of educational resources designed to support education for sustainable development (Department of Environment Transport and Regions, 1999). The code of practice has ten principles for production of teaching/information materials:

- Principles of sustainable development
- Integrity
- Balance
- Values and attitudes
- Knowledge and skills
- User-centered approach
- Need
- Development
- Production
- Promotion and distribution

It is interesting to note that user-centered approaches are one of the key ten principles. Here the notion of user-centered approaches is to ensure maximum adoption, so resources should be easy to use and appropriate to the intended audience as well as allowing flexibility in teaching and learning styles.

FIELDWORK TEACHING: ARE LEARNER-CENTERED APPROACHES POSSIBLE?

Fieldwork is *the* authentic context for teaching ecology, and certainly in the early days of ecology fieldwork was an integral part of scholarly wanderings and teaching. When student numbers were few, fieldwork teaching was logistically possible and academically made good sense. It relied on an expert, enthusiastic instructor providing guidance to students in an often unfamiliar but real environment, but it was rarely student-centered. The empirical approach to fieldwork is where ecology makes sense, and fieldwork as a teaching strategy lends itself well to student-centered approaches. Indeed, this approach is emphatically desirable if we are to avoid destruction of or damage to natural habitats through trampling by a class of fifty students! In other sciences such as physics and chemistry we see a similar emphasis on authentic contexts and settings, particularly for practical work, with student-centered approaches (Roth and Tobin, 2002). However, in these disciplines this emphasis has been retained as an essential component of higher education study, while in ecology it has weakened. Despite the fact that Professor John Grace FRSE (Fellow of Royal Society of Edinburgh), past president of the British Ecological Society and professor of environmental science at the University of Edinburgh, indicates that although "fieldwork at university is an essential part of an environmental scientist's training and fieldwork at school will be their inspiration" (p. 2), fieldwork is seriously under threat in all education sectors (quoted in Barker, Slingsby, and Tilling, 2002).

The causes of the decline in fieldwork are multifaceted, complex, and interrelated. They include cost, time, health and safety, availability of suitable sites, commitment and expertise of staff, and so on. Learner-centered approaches in fieldwork can overcome some of these difficulties as well as assist in the logistical problem of supporting students who may be widely distributed outside the confines of a lecture theater. Dashing from one group of students to another on a regular basis to provide guidance on plant or animal identification can result in "fieldwork fatigue." Developing interpretative trails for fieldwork can prevent this. Interpretative trails are widely used in informal educational settings, such as nature centers or museums, but very much underused in formal education, particularly higher education. Essentially, interpretative trails are self-guiding trails that allow learners to follow a trail made up of a series of experiences, such as observation of specific phenomena, tasks, or investigations. For example, in a local woodland, a temporary ecological trail was set up with ten stations in sequence marked by ranging poles with numbered flags; each station had a specific ecological student-centered task. The students worked in small groups, each group instructed to start at a different point on the trail and then guided by the teaching materials, which had been prepared in a style commensurate with an open learning style after Race (1989). The materials thus support student-directed learning.

Some of the criticisms of ecology highlighted in the Introduction can be a benefit in a teaching context, as we do not need high-tech, expensive equipment (di Castri and Hadley, 1986). This means that students can either take simple equipment with them or some equipment can safely be left at the relevant station on the trail. From a teaching point of view, students can fully participate in the ecological investigations without the physical presence of the instructor, thus the trail is very much a learner-centered experience. Moreover, student experiences are more sensibly shared in the plenary as there is a commonality of experience. The students will have looked at and investigated the same phenomena so the context of the discussion is the same and direct comparisons can be made. This approach differs from traditional fieldwork, where groups of students might have been observing or investigating the same phenomenon but in slightly different localities—which because of the inherent complexity of the natural world can be widely different. The commonality of experience on the trail and shared discussion of experiences leads to a much deeper understanding of the ecological concepts. The ability to share experiences in small group discussions while on task (at the students' own pace) and in a larger forum in the plenary is a vital component of meaningful learning in this context (Vygotsky, 1978).

If such a trail is used year after year, then the ecological data collected by the students can be built up into a meaningful long-term data set. The data are meaningful because the same location is used for each investigation and are thus both valid and reliable. Moreover, the ecological investigations become more meaningful to the student, as ecological change is a long-term process. Most ecological investigations usually provide only a snapshot image of the situation, and to appreciate dynamic processes in natural habitats, long-term investigations are essential. These, however, are impossible in traditional three-year undergraduate programs and thus the placing of a students' experience into a longer-term data set has clear benefits. The students are part of a larger community of learners, each cohort contributing to and benefiting from the data collected by others. The use of interpretative trails in higher education has been a pedagogical shift that, although expensive in terms of development time, has resulted in a more beneficial learning experience. In subsequent years there will be a payoff in terms of preparation time, although student-centered learning such as this still needs to be managed and facilitated, so it is by no means an easy option.

Virtual Field Trips

A newly emerging teaching tool is the use of virtual field trips (VFT), and while such an approach does represent an authentic context, the experience can be valuable in making better use of time in an actual field course, i.e., as a preparatory activity, or it can make an important contribution to learning where real fieldwork is not possible for reasons of cost, disability, danger, etc. It also has the advantage of being learner-centered! One only has to search for "virtual field trips" in a search engine on the World Wide Web to see a large range of initiatives in this area. In one initiative in the United Kingdom, virtual field trips have been designed as part of the LTSN-GEES-funded project "Atmosphere, Lithosphere, Hydrosphere, Biosphere: Cross-Disciplinary Virtual Fieldwork" <www.brookes.ac.uk/bms/vfw/>. VFT uses virtual quadrats to quantify the distribution of plants. This project uses chalk grassland, a widespread community that students encounter in local field studies. Each plant species is associated with an image and Web page including characteristics useful for its identification. This virtual exercise simulates the activity normally carried out in the field, where a sampling quadrat is thrown at random, in this case in an area of chalk grassland, and a count taken of the plant species that fall within it. This can be repeated several times to gather data that will be different every time but still reflects the characteristics of this particular area. The data can be viewed as a Web page or as a spreadsheet, and it can be saved for use in other applications.

No doubt the virtual fieldtrip is valuable, but when it comes to environmental responsibility, a growing body of evidence suggests that outdoor fieldwork experiences are the answer to bridging the gap between environmental knowledge and pro-environmental behavior (Kollmuss and Agyeman, 2002; Bogner, 1998). Indeed, Hillman (2002) has indicated that appreciation of beauty and the aesthetic qualities of the environment are the answer to taking onboard the premise of the precautionary principle. Dovers, Norton, and Handmer (1996) state that in cases of threats of serious or irreversible environmental damage, lack of scientific certainty should not be used as a reason for postponing measures to prevent environmental degradation. Hillman's thesis is that in today's psychology of a society in haste, it is not in human nature to exert caution and one can do so only by stopping, looking, and listening, the subsequent aesthetic human response to beauty will be a political action. The nature of fieldwork means that one does have to stop, look, and listen, and appreciation of the environment does indeed follow. Chawla (1999), for example, highlights such direct experience of nature as being a critical factor in determining the pro-environmental commitment of professional environmentalists in the United States. Moreover, Manzanal, Barreiro, and Jimenez (1999) present convincing evidence for the environmental benefits of fieldwork: "fieldwork helps clarify ecological concepts and intervenes directly in the development of the more favourable attitudes in defence of the ecosystem" (p. 431). These authors thus make a strong case for the benefits of direct interaction with the living world through fieldwork.

LEARNING INDICATORS

One of the problems involved in student-centered activities, particularly in fieldwork, is assessing whether learning is actually taking place when students are on task. The natural environment is an informal environment and there has been little published on indicators of learning in this context, although there is a growing body of evidence on learning indicators in other informal environments, such as museums (Griffin, 1999). In order to investigate this question, one could look at learning outcomes and/or the presence of learning processes or behaviors during fieldwork. A major dilemma is the difficulty in isolating and measuring cognitive learning outcomes for what is often a short-term experience, and in fact to attempt to do so conflicts with the constructivist learning paradigm, which describes learning as a developmental process involving the accommodation of new experiences with prior understandings and attitudes (Griffin, 1999). Fieldwork, although part of formal education, can be regarded as taking place in an informal setting where learning is often intrinsically motivated and proceeds through curiosity, observation, and activity (Ramey-Gassert, Walberg, and Walberg, 1994).

The motivation that comes from being in a novel environment means that the learning processes may thus be different in many respects from those normally associated with formal learning settings. A special opportunity offered by fieldwork is experiential learning, based on encounters with real organisms in authentic settings. It is a process that involves looking, questioning, examining, and comparing, and is rather similar to that described by Sheppard (1993) on a field trip. In informal settings such as fieldwork, cognitive and affective learning are fused and education and enjoyment are linked; thus it is less difficult for instructors to motivate students! Thus to evaluate whether learning is taking place, examination of the processes which indicate that learning is taking place is probably more effective than examining achievement of objectives (Griffin, 1999).

In their synthesis of literature on learning in museums Borun, Chambers, and Cleghorn (1996) list a number of behaviors related to learning which can be used as useful indicators of learning processes. These have been adapted into tools for assessing learning in fieldwork:

- Asking and answering questions
- Talking about a natural feature
- Pointing to a natural feature
- Reading resource materials
- Engaging in tasks
- Even "gazing" at the environment (adapted from Borun, Chambers, and Cleghorn, 1996, p. 135)

The fact that our students are adults means that when it comes to learning they have specific needs and motivations. Griffin (1999) summarizes the work of Knowles (1993) and Matthew (1996) who provide descriptions of adult learning. They note that adult students

- need to know why they should be learning something;
- have a deep need to be self-directing;
- incorporate their experiences as an integral part of them, they process new knowledge by reviewing it in the light of their experience;
- are ready to learn when their life experience provides a need to know;
- enter into a learning experience with a task-centered orientation to learning;
- are motivated to learn by both extrinsic and intrinsic motivations; and
- bring a whole agenda with them to learning situations.

It is reassuring to note that the need to self-direct is high on the agenda and this is reflected in the work of Evans and Abbott (1998) in their study of views of teaching and learning in higher education.

CONCLUSION

At a time of rapidly changing agendas in higher education, where teaching quality is being given more emphasis, there is a real need to evaluate teaching to ensure the output of highly skilled graduates who have the competencies and knowledge to act as responsible citizens. Indeed the U.K. subject benchmark statements produced by the Quality Assurance Agency for Higher Education (QAA) <www.qaa. ac.uk> serve as external reference points, identify expected skills and knowledge, and allow both external and internal evaluation of undergraduate programs as well as positively encouraging innovation in teaching. In the *Biosciences Benchmarking Statement* (QAA, 2002), student dialogue and student-centered teaching are highlighted, but little emphasis is given to fieldwork or the intangible benefits derived from student-centered activities, notably students becoming responsible citizens living a sustainable lifestyle. Such added value of student-centered approaches to learning in ecology needs to be supported by more empirical research and given more prominence. This chapter has presented a range of strategies to demonstrate that from a teaching, learning, and personal point of view student-centered learning in ecology is definitely worth the effort.

REFERENCES

Adams, D.M. (1989). Experience, reality and computer controlled technology. In R. Kraft and M. Sakofs (Eds.), *The theory of experiential education* (Second edition) (pp. 204-208). Boulder, CO: Association for Experiential Education.

Barker, S., Slingsby, D., and Tilling, S. (2002). *Teaching biology outside the classroom: Is it heading for extinction? A report on outdoor biology teaching in the 14-19 curriculum.* Shrewsbury, UK: Field Studies Council.

Bogner, F. (1998). The influence of short-term outdoor ecology education on long-term environmental variables of environmental perception. *Journal of Environmental Education, 29:* 17-29.

Borun, M., Chambers, M., and Cleghorn, A. (1996). Families are learning in science museums. *Curator, 39*(2): 123-138.

Bruner, J.T. (1993). *Schools of thought: A science of learning in the classroom.* Cambridge, MA: MIT Press.

Carson, R. (1962). *Silent spring.* Cambridge, MA: Riverside Press.

Chawla, L. (1999). Life paths into effective environmental action. *Journal of Environmental Education,* 31(1): 15-26.

Cobern, W.W. (1993). College students' conceptualisations of nature: An interpretive worldview analysis. *Journal of Research in Science Teaching,* 30(8): 935-951.

Cobern, W.W. (2000). *Everyday thoughts about nature: A worldview investigation of important concepts students use to make sense of nature with specific attention to science.* Dordrecht, the Netherlands: Kluwer Academic Publishers.

Department of Environment Transport and Regions (1999). *A voluntary code of practice: Supporting sustainable development through educational resources.* London: DETR.

di Castri, F. and Hadley, M. (1986). Enhancing the credibility of ecology: Is interdisciplinary research for land-use planning useful? *Geojournal,* 13: 299-325.

Dovers, S.R., Norton, T.W., and Handmer, J.W. (1996). Uncertainty, ecology, sustainability and policy. *Biodiversity and Conservation,* 5(10): 1143-1167.

Duschl, R.A. and Gitoma, D.H. (1991). Epistemological perspectives on conceptual change: Implications for educational practice. *Journal of Research in Science Teaching,* 28: 839-858.

Evans, L. and Abbott, I. (1998). *Teaching and learning in higher education.* London: Cassell Education.

Fensham, P.J., Gunstone, R.F., and White, R.T. (Eds.) (1994). *The content of science: A constructivist approach to its teaching and learning.* London: Falmer.

Griffin, J. (1999). Finding evidence of learning in museum settings. In E. Scanlon, E. Whitelegg, and S. Yates (Eds.), *Communicating science contexts and channels* (pp. 110-119). London: Routledge.

Hillman, J. (2002). The virtues of caution. *Resurgence,* 213: 6-9.

Knowles, M. (1993). Andragogy. In Z.W. Collins (Ed.), *Museums, adults, and the humanities* (pp. 49-60). Washington, DC: American Association of Museums.

Kolb, D. (1984). *Experiential learning: Experience as the source of learning and development.* Upper Saddle River, NJ: Prentice Hall.

Kollmuss, A. and Agyeman, J. (2002). Mind the gap: Why do people act environmentally and what are the barriers to pro-environmental behaviour? *Environmental Education Research,* 8(3): 239-256.

Kraft, R.J. and Kielsmeier, J. (1995). *Experiential learning in schools and higher education.* Boulder, CO: Association for Experiential Education.

Lambert J.M. (Ed.) (1966). *The teaching of ecology.* British Ecological Symposium No. 7. Oxford: Blackwell Scientific.

Manzanal, R.F., Barreiro, R., and Jimenez, M.C. (1999). Relationship between ecology fieldwork and student attitudes toward environmental protection. *Journal of Research in Science Teaching,* 36: 431-453.

Matthew, M. (1996). Adult learners. In G. Durbin (Ed.), *Developing museum exhibitions for lifelong learning* (pp. 70-72). London: Museums and Galleries Commission.

McConnell, D. (2000). *Implementing computer-supported cooperative learning* (Second edition). London: Kogan Page.

McIntosh, R.P. (1985). *The background of ecology concept and theory.* Cambridge, MA: Harvard University Press.

Orion, N. and Hofstein, A. (1994). Factors that influence learning during a scientific field trip in a natural environment. *Journal of Research in Science Teaching,* 31: 1097-1119.

QAA (2002). *Biosciences benchmarking statement.* Gloucester, UK: Quality Assurance Agency for Higher Education.

Race, P. (1989). *The open learning handbook: Promoting quality in designing and delivering flexible learning* (Second edition). London: Kogan Page.

Race, P. (1998). *500 tips on open and flexible learning.* London: Kogan Page.

Ramey-Gassert, L., Walberg, H.J.I., and Walberg, H.J. (1994). Re-examining connections: Museums as science learning environments. *Science Education,* 78(4): 345-363.

Roth, W. and Tobin, K. (2002). College physics teaching: From boundary work to border crossing and community building. In P. Taylor, P. Gilmer, and K. Tobin (Eds.), *Transforming undergraduate science teaching* (pp. 145-174). New York: Peter Lang Publishing Inc.

Salmon, G. (2000). *E-moderating: The key to teaching and learning online.* London: Kogan Page.

Sheppard, B. (1993). Aspects of a successful field trip. In B. Sheppard (Ed.), *Building museum and school partnerships* (pp. 38-52). Washington, DC: American Association of Museums.

Taylor, P. (1993). Collaborating to reconstruct teaching: The influence of researching beliefs. In K. Tobin (Ed.), *The practice of constructivism in science education* (pp. 267-297). Washington DC: AAAS Press.

Toyne, P. (1993). *Environmental responsibility: An agenda for further and higher education.* London: Her Majesty's Stationery Office, Department for Education.

Vygotsky, L.S. (1978). *Mind in society.* Cambridge, MA: Harvard University Press.

Wali, M.K. (1999). Ecology today: Beyond the bounds of science. *Nature and Resources,* 35(2): 38-50.

Chapter 2

The Use of ICT in Molecular Science Student-Centered Learning: A Developmental Approach

Philip L. R. Bonner

INTRODUCTION

Higher education (HE) in the United Kingdom has seen many changes during the past decade. Central to these changes has been the movement to lifelong learning in the learning age (Dearing, 1997; Department for Education and Employment [DfEE], 1998). This includes a change of emphasis in education delivery, away from didactic teaching of students to promoting independent student learning. Information and communication technology (ICT) is required to play a key role in this learning process, providing learners of all ages access to a worldwide knowledge base (Selwyn, Gorand, and Williams, 2001).

The expansion of the HE sector resulted in the development of new courses to accommodate the increasing student numbers but without the concomitant increase in resources. This increase in student numbers coincided with the advent of the modern personal computer, and it was inevitable that the HE sector would endeavor to utilize ICT to offset some of the resource deficiencies (Holt and Thompson, 1995). In addition, some institutions assume that the modern student finds the use of ICT appealing, but this assumption appears to be without good reason (Carpenter and Tait, 2001).

The majority of early software required little change in teaching practice and was designed to support rote teaching. The lack of good software was addressed in 1992 by the Teaching and Learning Technology Program (TLTP), which funded the development of educational software with a greater emphasis on the interaction of the learner with the computer program. Good educational software should be interactive, requiring that the learners participate rather than just watch (Schank, 1993). A vast range of different ICT applications are available, covering all areas of tertiary-level science teaching. Information (lists, reviews, and contact details) on the ICT applications can be found by visiting HE learning support Web sites (e.g., Learning and Teaching Support Network [LTSN] Biosciences, Leeds University, United Kingdom; Uniserve Science, The University of Sydney, Australia) or by using a Web-based search engine.

ICT AND STUDENT LEARNING

In general, ICT (multimedia compact discs [CDs], databases, video streaming, the Internet, virtual discussion groups, virtual labs, virtual learning environments, and computer-aided learning [CAL]) can be divided into two categories: Type 1 applications involve using computer resources to carry out teaching and learning processes that could be conducted without the need for a computer. Type 1 applications do not necessarily require a change in teaching strategies, but they can assist in independent learning and reduce the need for direct teacher involvement. Type 2 ICT applications involve using computers to carry out the teaching and learning processes in a fashion that had not been possible using conventional teaching methods (Lockhard, Abrams, and Many, 1994).

It has been proposed that, through the use of ICT, learners will acquire the necessary skills of observation, inquiry, and interpretation (Hawkey, 2002). Participation by students in the learning process has been enhanced by the rapid development of the World Wide Web (Tait, 2000). Learning is enhanced by the flexibility of the Internet, which allows the learner to control the pace of the learning process by selecting what, how long, and how many times to view educational material. In addition, the learner can select from many possible pathways, exploring different solutions to problems (Lin and Hsieh, 2001). In effect, learners can tailor the experience to suit their learning styles and be encouraged to become responsible for their own learning (Peat and Fernandez, 2000). These are part of the processes that Garrigan (1997) cites as facilitating effective student-centered learning. In addition, Long, Pence, and Zeilinski (1995) argue that the use of computer-based instruction enhances the reflective process and long-term learning. However, some disadvantages are present with Web-based delivery of multimedia modules; these include the use of non-peer-reviewed material, the difficulty learners may encounter in finding relevant information, and reducing the contact an instructor has with the student (Mudge, 1999). Indeed, access to ICT does not by itself encourage people to learn (Selwyn and Gorand, 2003).

Type 1 and Type 2 ICT: Blurring of the Boundary

Universities have rapidly embraced Web technology because Web-based teaching has the advantages of accessibility, ease of use, flexibility, and interactivity. Virtual learning environments (VLE) are increasingly being used to support online learning in HE institutions. The VLEs can be used as a Type 1 ICT by allowing the instructor to provide material to the learner that could be provided without the use of a computer. However, the VLE makes accessing educational material effortless—the instructor can upload word-processed documents, presentations, and Web files in a user-friendly environment (Kennewell, 2003). The VLE can also be used for communication between the learner and the instructor and for online assessments, which can provide immediate feedback to the learner (Peat and Franklin, 2002). When interacting with a VLE the learner may at first encounter a Type 1 ICT application, but by interaction with the VLE can be encouraged to become an active learner and in doing so encounter Type 2 ICT learning resources, including virtual laboratories and CAL. These resources should form the starting point for exploration, encouraging learner autonomy (see Chapter 1 by Barker in this book for further discussion on the use of a VLE to support learning).

Educational multimedia CDs provide an example of both Type 1 and Type 2 ICT applications. Publishers of molecular sciences textbooks now routinely include multimedia CDs as a component of the latest textbooks. In previous years, these CDs were little more than a collection of the two-dimentional images already present in the accompanying textbook. This has done little to promote a positive attitude toward computers or the subject matter, because learners experience discouragement and negative attitudes toward ICT when time is wasted on low-quality, irrelevant material (Breen et al., 2001). Recently, a selection of the latest editions of molecular science text-books has included multimedia CDs and associated Web sites (Alberts et al., 2002; Berg, Tymoczko, and Stryer, 2002). The material contained within the CDs and Web sites complements rather than duplicates the material in the textbooks and includes a number of items designed to enhance the learning experience. In addition, to the text and two-dimentional images (Type 1 ICT) there are interactive three-dimensional graphics (Type 2 ICT), which encourage the learner to observe and reflect upon the molecular structures. Two major strengths of the virtual structures are the ability to visually present what cannot normally be seen and to immerse the learner in that environment (Trindale, Fiolhais, and Almeida, 2002).

Optimistic Rhetoric?

Despite the enthusiastic adoption of ICT by the HE sector, documentary evidence that ICT has indeed any value as a learning aid, particularly in the bioscience area, is limited (Fox and Hermann, 1998). The limitations of using multimedia in HE have been highlighted (Davies and Crowther, 1995; Hughes and Daykin, 2002) and questions have been raised as to the value of simulated laboratories, given the fact that they are not "real" (Baggot la Velle, 2002).

Reynolds, Treharne, and Tripp (2003) also have raised doubts about some of the published academic research that extols the effectiveness of ICT in education. They argue that some of the published work is flawed and propose to "rein in" the optimistic rhetoric and conduct more research on improving ICT provision.

Indeed, a major problem is assessing the effectiveness of the ICT in replacing traditional teaching methods. Qualitative analysis and feedback from students can be used to assess the effectiveness of ICT (e.g., Wang, 2001; Franklin, Peat, and Lewis, 2002; Shim et al., 2003), but this method of evaluation is open to the criticisms raised against qualitative analysis, i.e., small sample sizes, anecdotal comments, flawed questionnaires, and enthusiastic instructors biasing the group's response (Clark, 1983). However, given the complexities of the many interactions in such experiments, it is not surprising that qualitative analysis is chosen. Quantitative analysis of the effectiveness of ICT material in helping students to learn is problematic because it is best used to answer specific questions. A straightforward method to generate quantitative data is to replace only one element of traditional teaching practices at a time and then assess the effectiveness of the change in the learning outcomes (e.g., Bonner, 2000; Traver et al., 2001; Green, 2002). A period of reflection is then required, followed by modification to the methodology and another period of evaluation. The net result is a gradual development of the effective use of ICT in student learning rather than unquantifiable wholesale change.

The diversity of biosciences and the large number of ICT packages available precludes any meaningful discussion of all of them, so it is the intent of this chapter to demonstrate a developmental approach to the use of ICT in supporting molecular science student-centered learning. As such it can be considered a guide rather than a blueprint.

CAL: A LEARNER-CENTERED APPROACH
TO NUMERACY PROBLEMS

Students enrolling in biological science courses at Nottingham Trent University (NTU) arrive with an educationally varied background, many with standard entry qualifications but some with non-traditional entry qualifications. The increase in student numbers has inevitably led to a decrease in personnel contact time with individual students, and these bioscience students, like other science students, seem to learn without understanding the concepts. This is probably due to many different factors, but one that has been cited is the magnitude of the course content, which allows little time for reflection (Fox and Radloff, 1998; Cann and Seale, 1999). The use of ICT can encourage reflection as part of the learning process (Long, Pencer, and Zeilinski, 1995), but ICT cannot completely replace instructor-supported learning, which is valued by students (Oliver and Omari, 1999).

In line with other U.K. HE institutions, at NTU we have noticed a decline in numeracy skills among first-year students admitted to our bioscience courses (Tariq, 2002). This lack of numeracy is manifest in the students' lack of ability to manipulate numbers, a failure to correctly convert between units (also a lack of appreciation of the importance of units to science), and an inability to use basic equations.

These numerical deficiencies are addressed by using CAL packages such as the following.

1. *Pharmacy Consortium Computer Aided Learning (PCCAL): Basic Calculations in Pharmacy*—This is a tutorial on the theory and methodology of basic calculations in biosciences. The CAL programs from the PCCAL group are available as 32-bit Windows- and Web-based packages, and there are over eighty individual packages covering a diverse range of topics from laboratory safety to molecular biology. The PCCAL programs are designed for student-centered learning, providing an interactive tutorial on the subject matter that allows self-paced progression.

2. *Question Mark Perception*—This is a Web-based assessment program that allows the instructor to author self-assessment tests, which can be delivered online. When the self-assessment tests are accessed by the learner, the program randomly generates a unique test from a bank of questions, providing the learner with instant feedback on his or her improvement in numerical manipulations. Both programs can be accessed at times convenient to the learner and remain accessible throughout the duration of the course, providing continuity in support for numerical calculation deficiencies.

EVOLVING EVALUATION

Many students adopt a pragmatic approach to learning. Elton (1996) suggests that students are motivated to learn in part because of assessment. Therefore, to improve the students' commitment to learn, it is important to make the students feel confident that they will be able to pass the examination. To evaluate the effectiveness of and to motivate the students' access to the CAL programs, students are required to take a pre-CAL usage numerical test and, after four weeks' exposure to the CAL program, they retake the test which counts toward their module mark. After the first assessment the students undertake self-diagnosis of their numeracy problems; they are then given clear instructions on how they can address the problems and what the end benefit will be. Table 2.1 shows the results of using the

TABLE 2.1. Percentage Results of Biochemical Numerical Tests Before and After the Use of CAL Programs [a]

	1998, HND year 1 ($n = 48$)	2000, HND year 1 ($n = 37$)	2001, HND year 1 ($n = 27$)	2002, UFD year 2 ($n = 17$)	2003, UFD year 2 ($n = 17$)
Test A[b]	48 ± 14	0.3 ± 0.5	0.4 ± 0.5	8.5 ± 16.3	0.0
Test B[c]	73 ± 21	38 ± 32.7	50.1 ± 28.2	66.4 ± 30.4	57.1 ± 34.3
Test C[d]	–	–	–	–	76.5 ± 24.3

HND = higher national diploma; UFD = university foundation degree.
[a]Mean ± standard deviation; [b] Test A before the use of the CAL programs; [c] Test B four weeks after using the CAL programs; [d] Test C two weeks after Test B using the CAL programs.

CAL programs with a group of higher national diploma (HND)/university foundation degree (UFD) students over a five-year period. Over the five-year period, the CAL programs have remained constant but the evaluation has evolved. In 1998 the students undertook a multiple-choice question (MCQ) pre– and post–CAL usage test that was capable of masking the true extent of their numerical deficiencies. In 2000 and subsequent years, a written test was used that revealed the true level of numerical deficiency but also highlighted how effective the CAL programs can be in promoting student learning. In 2003 an additional test was included six weeks after the initial test to try and ascertain the retention of the knowledge acquired.

After analysis of the pre– and post–CAL usage tests during periods of reflection, it is possible to make minor modifications to the evaluation process. However, as only one component is altered, the results can be viewed with confidence. In this case the results (Table 2.1) collectively indicate that the CAL programs were capable of facilitating student learning. Student comments about using the CAL programs were in the main positive, and typical comments can be seen in Box 2.1. The programs did not suit the learning styles of all students, but with careful tutoring and support in subsequent years, the students valued the accessibility of programs throughout the course and the provision of support for their numeracy problems.

BOX 2.1. Student Comments About the Use of the CAL Numeracy Programs, 1998 ($n = 30$)

Computer usage, maximum of one time (n = 12)

- "The Course has improved my knowledge of units and concentration by making me reconsider them. This has reinforced my knowledge."
- "I feel my knowledge of units and concentration has improved."
- "Tutorial lessons should be held in the computer rooms, otherwise you put it off and don't use it to the fullest."
- "Unsure of which program to use."

Computer usage, two to five times (n = 28)

- "The package is very good, and it is a good way to learn."
- "I do believe my knowledge has improved since using the computer program."
- "The computer programs were helpful in improving my knowledge of units and concentration, although it wasn't vital to use it."
- "My knowledge has improved, as the program explained the subject quite clearly."

A LEARNER-CENTERED APPROACH TO A LABORATORY SIMULATION

Practical laboratory skills are the critical skills that need to be acquired by students in biosciences. However, some areas of biosciences do not easily lend themselves to laboratory exercises that develop these skills. One such area is protein purification, in which the process often requires multiple steps conducted over many days. Prior knowledge of the properties of proteins and the techniques that can be used is necessary, as well as the ability to formulate a strategy to isolate an individual protein from a complex mixture. The equipment requirements, the increase in student numbers, and the time-scale of the purification process make it difficult to conduct a meaningful protein purification practical in limited laboratory time. Although a CAL package cannot replace the experience of laboratory work, it does allow students to experience, in a manageable time frame, the empirical nature of the subject matter without the resource requirements.

A CAL program, *protein*LAB (European Academic Software award [EASA] winner, 2002), is one of a number of programs from the eLABorate group (York University, United Kingdom) which are available in a 32-bit Windows- or Web-based format. All of the eLABorate packages have been designed to encourage interactivity by providing the opportunity to conduct a virtual experiment. The student is allowed to control the course of the experiment and interpret the data, and in doing so he or she experiences the empirical nature of scientific investigation in a compressed time frame without the cost of laboratory work (Garratt, 2000). The *protein*LAB program allows students to use different protein purification techniques, either in isolation or in combination, to construct a purification strategy under realistic conditions. Information on the various techniques is available in help files, but the success (purification of a target protein to homogeneity) or failure of the session is entirely attributable to the decisions made by the student.

Developing the Delivery Mode

At NTU, prior to using *protein*LAB, the topic of protein purification was delivered to the students through a series of lectures on several techniques, followed by a six-hour practical on the use of one technique. This failed to convey the empirical nature of protein purification because in the time allowed the students could not use the information to suggest improvements in the protocol. When *protein*LAB was first introduced to the teaching schedule, it was sandwiched between lectures on techniques and the laboratory. The delivery was crowded and again failed to let the students experience the empirical nature of the subject matter. The feedback from the students was positive to the computer package, with a request for more time to be allocated in using the computer package (Table 2.2).

As a consequence, the lectures and laboratories have been replaced by sessions using the CAL program. A "failure-driven learning approach" (Barnard and Sandberg, 1992) is used with the CAL program, providing the learner with support as the learner needs it rather

TABLE 2.2. Student Feedback Collected in 1998 and 2003 About the Use of *protein*LAB.

Year	Question	Response	Additional comments
1998	What has increased your knowledge of protein purification?	11/13 rated the CAL better than the lectures.	"The computer package was very useful."
	Do you think there should be more time on the CAL?	7/13 thought there should be more time using the CAL. 6/13 thought there was the correct amount of time on the CAL package.	"More time was required for the computer package, as only five out of the twenty proteins were done."
2003	Overall the simulations were . . .	Only 1/8 students rated the simulations poor.	"More lectures, more computer time."
	Has the computer package been of value to increase your knowledge of protein purification?	5/8 rated the simulations of great value.	"Most international students have different backgrounds. We never had such packages so it would be useful if you could start with some basics and more lectures."
	Do you think there should more or less?	2/8 students wanted less CAL than lectures. 6/8 students wanted more lectures and CAL.	

than in advance of need. Prior to contact with *protein*LAB, the students are given a printed handout detailing the theory of the techniques they will be required to use, a lecture on the strategy of protein purification (emphasizing the need to minimize the number of steps and maximize the yield), and a tutorial on the basic calculations associated with protein purification. They then undertake a number of tutored sessions that are halted at appropriate intervals in order to mentor students on the correct use of the next technique. Information on using the software for each technique is preceded by appropriate theory. Advice on how to proceed is proffered, but direct instructions on how to proceed are avoided. After the period of mentoring, the students work at their own pace, applying the theory to experimental design, data collection, and interpretation. They receive instant feedback because the results of their decisions are immediately displayed on the screen. Random progression through the package can produce results, but by using prior knowledge and the information provided by the results of each experiment, the student can begin to make educated choices on the purification strategy.

By interacting with *protein*LAB, the students experience hypothesis, receive immediate feedback, and are required to pose questions and manipulate objects, all of which aids knowledge construction (O'Loughlin, 1992). In this fashion, meaningful learning is promoted when new knowledge is integrated with a prior knowledge base (Novak, 2002). The proteins contained within the program can be purified using many different protocols, which allows students to discuss and compare strategies. This allows an element of self-assessment and evaluation that is a desirable constituent of the learning process (Stefani, 1998).

Evaluation

The delivery of protein purification at NTU has developed gradually from a didactic to a constructivist style. Qualitative feedback in Table 2.2 indicates that the majority of students enjoyed the experience of using the program as an alternative to the traditional teaching approach, and the exam marks for the students have remained constant. However, not all feedback was positive. A minority of students did not warm to the topic or the package and seemed to prefer traditional teaching practices. This feedback may reflect different learning styles, as Garratt (2000) reported similar feedback when he evaluated other eLAB packages and reflected that many students cannot apply their knowledge of the subject in a flexible and creative way, and thus may find the process demotivating. He suggested that careful preparation and debriefing by the instructor could avoid this problem. Again, the CAL material remains constant; what gradually changes is the delivery of the material by the instructor in response to students' needs and the evaluation of their learning.

CONCLUSION

Dearing (1997) has predicted a learning society, and part of the process is learning to learn (Rawson, 2000). The use of ICT can offer a highly interactive, student-centered learning experience, which contains elements that promote knowledge construction (O'Loughlin, 1992). In addition, the learning experience can be presented to students in a variety of ways, which is likely to appeal to individuals with different learning styles (Topping et al., 1996; Peat and Fernandez, 2000; Franklin, Peat, and Lewis, 2002; Smith, 2002). However, questions are being raised about the value of ICT in education because some published papers cannot justify the claims made (Reynolds, Treharne, and Tripp, 2003).

Changes to the structure of tertiary education, including the increase in student numbers, has reduced student contact time with lecturers, weakening the support available to students and presenting problems for the delivery of some course material. The educational goal of intellectual independence is not easily achieved with a large group, and this has forced instructors to adopt new teaching strategies. As a result there have been mixed reports on student attainment (Fletcher, 1999; Scott, Buchanan, and Haigh, 1997; Holt et al., 2001).

The increase in student numbers has reduced the available contact time with instructors, and this has meant that students need to become more proactive in the learning process rather than relying upon didactic teaching styles and rote learning. ICT is being promoted as a means to replace the loss of contact time with an instructor, particularly through the use of self-paced learning modules. However, ICT cannot be seen as a "panacea to cure all ills." The idea that ICT can totally replace contact with an instructor is far from the truth, as contact between students and instructors has been identified as an important element of the learning process (Garrigan, 1997; Hodson, Connolly, and Saunders, 2001). Student-centered learning in general cannot be achieved by semantics; learners must be given the support to learn (Garrigan, 1997; Steckler et al., 2001). To justify the use of ICT in student learning requires a coordinated, long-term, developmental approach that is planned and integrated with pedagogy. However, the increase in student numbers and a push to reduce costs will dictate that ICT will continue to play an increasingly important role in the education of students. Computing power will continue to be upgraded so we can look forward with anticipation to the next generation of more effective interactive educational technologies to support student learning.

REFERENCES

Alberts, B., Johnson, A., Lewis, J., Ralf, M., Roberts, K., and Walter, P. (2002). *Molecular biology of the cell* (Fourth edition). New York: Garland Science.

Baggott la Velle, L. (2002). "Virtual" teaching, real learning? *Journal of Biological Education,* 36: 56-57.

Barnard, Y. and Sandberg, J. (1992). Interviews on AI and education: Beverly Woolf and Roger Shank. *AICom,* 5: 148-155.

Berg, J.M., Tymoczko, J.L., and Stryer, L. (2002). *Biochemistry* (Fifth edition). New York: WH Freeman and Co.

Bonner, P.L.R. (2000). Assessment of the use of CAL to replace remedial biochemical calculation tutorials. *CAL-laborate,* 6: 8-9.

Breen, R., Lindsay, R., Jenkins, A., and Smith P. (2001). The role of information and communication technologies in a university learning environment. *Studies in Higher Education,* 26: 95-114.

Cann, A. and Seale, J. (1999). Using computer-based tutorials to encourage reflection. *Journal of Biological Education,* 33: 130-132.

Carpenter, B. and Tait, G. (2001). The rhetoric and reality of good teaching: A case study across three faculties at the Queensland University of Technology. *Higher Education,* 42: 191-203.

Clark, R.E. (1983). Reconsidering research on learning from media. *Review Journal of Educational Research,* 53: 445-449.

Davies, M.J. and Crowther, D.E.A. (1995). The benefits of using multimedia in higher education: Myths and realities. *Active Learning,* 3: 3-6.

Dearing, R. (1997). *Higher education in the learning society: Report of the National Committee of Inquiry into Higher Education.* London: NCIHE.

Department for Education and Employment (DfEE) (1998). *The learning age: A renaissance for a new Britain.* London: DfEE.

Elton, L. (1996). Strategies to enhance student motivation: A conceptual analysis. *Studies in Higher Education,* 21: 57-68.

Fletcher, H.L. (1999). Reduction in attainment with increasing student numbers in a university first level genetics class. *Journal of Biological Education,* 34: 32-35.

Fox, R. and Hermann, A. (1998). Communicating through computers: Changing teaching in changing times? In C. Rust (Ed.), *Improving student learning—Improving students as learners* (pp. 558-564). Oxford: Oxonian Rewley Press.

Fox, R. and Radloff, A. (1998). No time for students to learn important skills? Try "unstuffing" the curriculum. In C. Rust (Ed.), *Improving student learning—Improving students as learners* (pp. 565-572). Oxford: Oxonian Rewley Press.

Franklin, S., Peat, M., and Lewis, A. (2002). Traditional versus computer-based dissections in enhancing learning in a tertiary setting: A student perspective. *Journal of Biological Education,* 36: 124-129.

Garratt, J. (2000). Simulating biochemistry: The eLABorate project. *CAL-laborate,* 4: 6-10.

Garrigan, P. (1997). Facilitating effective student centered learning: Enablement and ennoblement. *Journal of Further and Higher Education,* 21: 97-105.

Green, J. (2002). Replacing lectures by text-based flexible learning: Students' performance and perceptions. *Journal of Biological Education,* 36: 176-180.

Hawkey, R. (2002). The lifelong learning game: Season ticket or free transfer? *Computers and Education,* 38: 5-20.

Hodson, P., Connolly, M., and Saunders, D. (2001). Can computer-based learning support adult learners? *Journal of Further and Higher Education,* 25: 323-335.

Holt, D. and Thompson, D. (1995). Responding to the technological imperative: The experiences of an open and distance education institution. *Distance Education,* 16: 43-64.

Holt, R.I., Miklaszewicz, P., Cranston, I.C., Russell-Jones, D., Rees, P.J., and Sonksen, P.H. (2001). Computer-assisted learning is an effective way of teaching endocrinology. *Clinical Endocrinology,* 55: 537-542.

Hughes, M. and Daykin, N. (2002). Toward constructivism: Investigating students' perceptions and learning as a result of using an online environment. *Innovations in Education and Technology International,* 39: 217-224.

Kennewell, S. (2003). Editorial: Real learning in a virtual environment. *COMPUTER Education,* 103: 1.

Lin, B. and Hsieh, C. (2001). Web-based teaching and learner control: A research review. *Computers in Education,* 37: 377-386.

Lockhard, J., Abrams, P.D., and Many, W.A. (1994). *Microcomputers for the twenty-first-century educators.* Reading, MA: Addison-Wesley.

Long, G., Pence, H., and Zeilinski, T.J. (1995). New tools vs. old methods: A description of CHEMCONF '93 discussion. *Computers in Education,* 24: 259-269.

Mudge, S.M. (1999). Delivering multimedia teaching modules via the Internet. *Innovations in Educational Technology International,* 36: 11-16.

Novak, J.D. (2002). Meaningful learning: The essential factor for conceptual change in limited or inappropriate propositional hierarchies leading to empowerment of learners. *Science Education,* 86: 548-571.

Oliver, R. and Omari, A. (1999). Using online technologies to support problem-based learning: Learners' responses and perceptions. *Australian Journal of Educational Technology,* 15: 58-79.

O'Loughlin, M. (1992). Rethinking science education: Beyond Piagetian constructivism toward a sociocultural model of teaching and learning. *Journal of Research in Science Teaching,* 29: 791-820.

Peat, M. and Fernandez, A. (2000). The role of information technology in biology education: An Australian perspective. *Journal of Biological Education,* 34: 69-73.

Peat, M., and Franklin, S. (2002). Supporting student learning: The use of computer-based formative assessment modules. *British Journal of Educational Technology,* 33: 515-523.

Rawson, M. (2000). Learning to learn: More than a skill set. *Studies in Higher Education,* 25: 225-238.

Reynolds, D., Treharne, D., and Tripp, H. (2003). ICT—The hopes and the reality. *The British Journal of Educational Technology,* 34: 151-167.

Schank, R.C. (1993). Learning via multimedia computers. *Technology in Education,* 36: 54-56.

Scott, J., Buchanan, J., and Haigh, N. (1997). Reflections on student-centered learning in a large class setting. *British Journal of Educational Technology,* 28: 19-30.

Selwyn, N. and Gorand, S. (2003). Reality bytes: Examining the rhetoric of widening educational participation via ICT. *British Journal of Educational Technology,* 34: 169-181.

Selwyn, N., Gorand, S., and Williams, S. (2001). The role of the technical fix in UK lifelong learning. *International Journal of Lifelong Learning,* 20: 255-271.

Shim, K.C., Park, J.S., Kim, H.S., Kim, J.H., Park, Y.C., and Ryu, H.I. (2003). Application of virtual reality technology in biology education. *Journal of Biological Education,* 37: 71-74.

Smith, J. (2002). Learning styles: Fashion, fad, or lever for change? The application of learning style theory to inclusive curriculum delivery. *Innovations in Education and Technology International,* 39: 63-70.

Steckler, A., Farel, A., Bontempi, J.B., Umble, K., Polhamus, B., and Trester, A. (2001). Can health professionals learn qualitative evaluation methods on the World Wide Web? A case example. *Health Education Research,* 16: 735-745.

Stefani, L.A. (1998). Assessment in partnership with learners. *Assessment and Evaluation in Higher Education,* 23: 339-350.

Tait, M. (2000). Visual media in computer-based learning. *Life Sciences Educational Computing,* 11: 5-11.

Tariq, V.N. (2002). A decline in numeracy skills among bioscience undergraduates. *Journal of Biological Education,* 36: 77-83.

Topping, K.J., Watson, G.A., Jarvis, R.J., and Hill. S. (1996). Same-year paired peer tutoring with first year undergraduates. *Teaching in Higher Education,* 1: 341-346.

Traver, H.A., Kalsher, M.J., Diwan, J.J., and Warden, J. (2001). Student reactions and learning: Evaluation of a biochemistry course that uses Web technology and student collaboration. *Biochemistry and Molecular Education,* 29: 50-53.

Trindale, T., Fiolhais, C., and Almeida, L. (2002). Science learning in virtual environments: A descriptive study. *British Journal of Education Technology,* 33: 471-488.

Wang, L. (2001). Computer-simulated pharmacology experiments for undergraduate pharmacy students: Experience from an Australian university. *Indian Journal of Pharmacology,* 33: 280-282.

USEFUL WEB SITES

Association of Clinical Biochemists CAL site
<http://www.acb.org.uk/>

eLAB
<http://www.york.ac.uk/depts/chem/staff/elaborate/>

European Academic Software awards
<http://www.easa-award.net>

Howard Hughes Medical Institute site
<www.biointeractive.org>

Learning and Teaching Support Network (LTSN bioscience)
<http://bio.ltsn.ac.uk/>

Lifesign Web site
<http://www.lifesign.ac.uk>

PCCAL
<http://www.coacs.com/PCCAL/>

proteinLAB
<http://www.york.ac.uk/depts/chem/staff/elaborate/packages/protein/protein.html>

Question mark perception
<http://www.questionmark.com>

Uniserve Science, University of Sydney, Australia
<http://science.uniserve.edu.au/>

Vega Science Trust
<http://www.vega.org.uk>

Chapter 3

The Undergraduate Life Sciences Laboratory: Student Expectations, Approaches to Learning, and Implications for Teaching

Janet Gorst
Susan Lee

Science teaching must take place in a laboratory; about that at least there is no controversy. Science simply belongs there as naturally as cooking belongs in a kitchen and gardening in a garden. (Solomon, 1994, p. 7)

INTRODUCTION

This quotation firmly places laboratory work at the core of science teaching, yet the teaching approaches to undergraduate laboratories often fail to convey the right messages or excite the students, thus wasting a valuable learning environment. When students do not even accept that they should understand purpose in laboratory work (Gunstone, 1991) it is time for serious reflection, because the way in which students respond to learning in the laboratory context depends, to some extent, on their perception of its purpose. This aspect, some-

We are especially grateful to Margaret Buckridge (lecturer, Griffith Institute for Higher Education) for her advice and encouragement during the time the project was carried out. A very big thank-you also to the staff and students of BBS3030, Biological Sciences Laboratory (particularly Angela Clark, Shannon Dillon, Damian Hatchett, Sam Lukowski, Beryl McDonald, Michelle Morley, Renee Norris, and Renee Stirling) who so generously gave us their time and their comments.

times referred to as the "hidden curriculum" (Snyder, 1971), is often overlooked by laboratory course designers, who tend to see a laboratory session in the context of what they planned rather than what is actually happening (Wilkinson and Ward, 1997).

The ways that the purpose and relevance of laboratory work are presented to students vary considerably between course designers, and this diversity in presentation may send out mixed messages even within a single university department. Although they are not direct pedagogical issues and could perhaps be termed "administrative overlays," there are two negatively impacting factors that must be mentioned when considering successful learning in the laboratory. First, to maintain and staff teaching laboratories is a significant expense for university departments, and laboratory programs are one of the main reasons for the cost differential between arts and science subjects. Financial constraints must be considered a major factor in the list of reasons for the frequent failure of undergraduate laboratory classes to deliver the outcomes we would like. Second, there is often a mind-set at departmental level that the job of designing and running laboratory classes should be left to junior staff. This is a terrible irony, given that these inexperienced staff are expected to coordinate budgets, equipment, technical staff, academics, and students to produce a meaningful educational experience. Neither of these factors is considered in detail in this chapter, but it must not be forgotten that they often direct the way in which the learning environment in the laboratory classes is structured.

Although there exists a strong literature on the philosophies behind and reasons for laboratory work, the comment has been made that "science laboratory work needs a solid research basis to provide effective learning" (Feteris, Gunstone, and Mills, 2001, p. 47). The evolution of teaching projects such as APCELL, The Australian Physical Chemistry Enhanced Laboratory Learning (Barrie et al., 2001, p. 23), which aims to develop "a protocol for the design of teaching experiments based upon sound pedagogical tenets," provides an encouraging example of nationally based attempts to overcome the perceived deficiencies in laboratory classes. A large project on undergraduate laboratory learning is also currently under way at Monash University in Australia and will track a cohort of physics students through their first, second, and third year (Feteris, Gunstone, and Mills, 2001). The first stage of the project has focused on surveys of first-year students to ascertain student perceptions of the purpose of school laboratory work, expectations of tertiary studies, experiences in the laboratory, and perceptions of the intention and value of individual laboratory activities. This valuable project will provide solid data that will assist in the development of better laboratory classes in all areas of science.

This chapter seeks to examine aspects of student-centered learning in the undergraduate life sciences laboratory. A study of third-year biology students at Griffith University in Australia is specifically used to illustrate that there is diversity in what students bring to laboratory work, what they perceive to be the reasons for and relevance of laboratory work, how they learn, and what learning difficulties they have.

A REVIEW OF THE LITERATURE

The Purpose of Laboratory Work

So what is it that we are trying to achieve in the laboratory? The importance of laboratory work in science teaching in Australia is emphasized by the Australian Education Council (1994) which believes that students should have the ability to "recognize and value scientific knowledge as reliable knowledge, based on observations, reproducible experiments and logic . . . and . . . evaluate experiments and arguments and the validity of results" (p. 84). These ideals often are not met (Wilkinson and Ward, 1997). Maienschein and students (1998) have defined two important areas in science learning, namely, scientific literacy and science literacy. Ideally, for those students intending to follow a career in science, it is essential that they have an integrated appreciation of the theory (science literacy) and that they learn something about the goals of trained scientists, the methods and procedures they use and the ways in which they communicate their results (scientific literacy). This scientific inquiry outcome, which integrates scientific and science literacy, is regarded as an important purpose of laboratory work (e.g., Boud, 1986). Hodson (1996) believes that it achieves three kinds of learning, namely, (1) enhanced conceptual understanding, (2) enhanced procedural knowledge, and (3) enhanced investigative expertise. The purposes that Hodson (1995) proposes for laboratory work reflect the broad categories of other science educators (e.g., Anderson, 1976; Hegarty-Hazel, 1990b) as follows:

- To motivate by stimulating interest and enjoyment
- To teach laboratory skills
- To enhance the learning of scientific knowledge
- To give insight into scientific method and to develop expertise in using it
- To develop certain "scientific attitudes," such as open-mindedness, objectivity, and willingness to suspend judgment

It is probably fair to say that the science laboratory experience at the undergraduate level does not closely reflect actual scientific practice. The administrative overlays mentioned earlier, such as large class sizes, limited equipment, and timetable scheduling often mean that the laboratory classes need to be run along strict lines that do not encourage scientific exploration. In particular, such laboratory classes are unable to teach the tacit knowledge at the "central core of the art and craft of the scientist" (Hodson, 1996, p. 130); this comes only through the experience of doing science. Hodson (1995) also refers to the "noise problem" encountered by students in a laboratory class. With each experiment the students are challenged not only by the problem being presented but also by issues such as how to set up and use the apparatus, how to collect and interpret the data, what outcomes to expect, and how to write up the experiment. With so much to think about, the students end up using poor pedagogical coping strategies such as copying others, following instructions without thought, focusing on only one aspect of the experiment, or, conversely, trying to cover everything randomly and being very busy getting nowhere.

Various studies suggest that laboratory work alone is not sufficient to bring about the breadth of conceptual development that we, as educators, wish for in undergraduate science laboratory classes. They also suggest that supplementary nonlaboratory activities may be beneficial. For example, Hodson (1995) acknowledges that the use of computer simulations can aid students' appreciation of scientific literacy (hypothesis generation and experimental design) to a degree not possible in traditional laboratory work, and Zadnik and Yeo (2001) have used to excellent effect a studio equipped to support student-centered, interactive, technology-based learning in the teaching of physics.

Pedagogical Issues

Several researchers (e.g., Staer, Goodrum, and Hackling, 1998; Feteris, Gunstone, and Mills, 2001) have used a scale of openness to inquiry elaborated by Hegarty-Hazel (1986) to classify laboratory activities. At the lowest level of inquiry (Level 0) the problem to be investigated, the apparatus to be used, the procedure, and the answer to the problem are all given to the students by the teacher or as a worksheet. Bain (1994) uses the terms *reproductive practice* and *transformative practice* to describe methods by which students learn (or are taught), and it is clear that Level 0 is based on reproductive practice, in which complex methods and ideas of a discipline are reduced to routine practice via convergent thinking. This represents the furthest pole away from student-centered learning. At the highest level of inquiry (Level 3) student-centered learning is at the core and students are required to determine all of the outcomes for themselves. In Bain's (1994) terminology, Level 3 is transformative and seeks to explore meaning through divergent thinking. However, while this type of learning is desirable, the use of more open inquiry is perceived as difficult by many teachers because of curriculum time constraints and behavior and safety management problems (Staer, Goodrum, and Hackling, 1998). These administrative overlay issues can impact negatively on the learning environment.

Integration of Theory and Practice

Of relevance to science learning is the balance between theory and practice. In 1962 Joseph Schwab, professor of education at Chicago University, developed an approach known as "inquiry methods" in science teaching. His aim with laboratory work was to give students a firsthand experience of the difficulties of specifying an actual problem and collecting data. He believed that experiment should precede classroom instruction and that the laboratory manual should not tell students what to do but, rather, should be written so as to encourage a student-centered focus on lateral thinking about where to find problems (i.e., transformative practice). Hodson (1996) maintains that the discovery learning that consequently developed in both England and America in the 1960s led to a highly distorted view of science and did not represent the way in which real scientists proceed. In particular he makes the point that it is not possible to discover something for which you are conceptually unprepared.

White (1991) places great emphasis on the ability of laboratory work to be a prolific source of "episodes" (unusual and striking incidents that form recollections of events). Students can directly link the episodes with the theoretical concepts gathered from more teacher-centered lessons to help in their understanding of that theory. Baird (1990) and Gunstone (1991) believe that the laboratory should train for enhanced student knowledge, awareness, and control of one's own learning, and that a vital component of learning is that students be able to link the concepts and experiences that they bring to a class with the practical work they undertake.

A LABORATORY STUDY

The Nature of the Study

In 2001, we undertook a project at Griffith University to look at the ways in which students in a third-year biology-based class learn laboratory skills. This project was linked to a study that one of us (JRG) was completing as part of an assessment item for a graduate certificate in higher education. In the process we wanted to consider some of the factors that Prosser and Trigwell (1999) highlight as being influential in learning through their presage-process-product model. This model sees perceptions of learning as an interaction between previous experiences of learning (presage), the learning context and the students' approach to learning (process), and the students' learning outcomes (product). Several factors were presented to students in questionnaires or through interviews. The factors fell into the following broad categories: the past experiences of laboratory work that the students bring to the subject; the perceived relevance of the subject; preparation for classes; learning expectations; enjoyment of the laboratory sessions; learning outcomes; approaches to learning; and specific preparation for the examination.

The Laboratory Course

Although the course was primarily laboratory-based (and this aspect was our only study focus) it also contained assessment items that exposed students to tasks allied to the laboratory work. The practical work, supplemented by a lecture/information session, consisted of eight laboratory sessions run once per week for four hours each week. The students, who worked in pairs, purchased a comprehensive set of laboratory notes that outlined the general tasks for each laboratory session and provided useful background information on specific techniques or items of equipment. Students were expected to have read the relevant preliminary notes and to have completed a number of computations/questions prior to each session; there were also computations and comments required at the end of each laboratory session, that usually were not due until the following week. Marks were awarded for this work. The practical work was designed to expose students to methodology that is widely used in biology (e.g., enzyme-linked immunosorbent assay [ELISA], sodium dodecyl sulphate-polyacrylamide gel electrophoresis [SDS-PAGE], immunoblotting, aseptic technique) as well as to a range of diagnostic techniques covering microbiology, immunology, haematology, clinical biochemistry, and pharmacology. The course was structured such that certain key aspects, e.g., ELISA, were repeated with variations over several weeks to reinforce learning. Also, several sessions required the students to design their own experimental testing of a problem, but with teaching staff being readily available for the students to discuss any queries. The nature of the practical work and practical assessment meant that there were requirements for mastery in areas such as dexterity in handling equipment to maximize accuracy, ability to troubleshoot experimental setups, an ability to interpret data, and the need to be able to apply knowledge to different situations. These attributes could not be rote learned and required the students to develop a critical awareness of what they were doing.

The Study

We took two main approaches to gathering information. First, we conducted two surveys of the whole class (sixty-six students enrolled). The first survey, conducted before the laboratory work began, probed the students' background, their attitudes about laboratory work, their expectations of the course, the perceived relevance of the course to their future aspirations, and how they would prepare for each laboratory class. The second survey, conducted in the final revision week, asked about the students' enjoyment of the course, what they had learned, how they learned practical work, their preparation for laboratory classes, and which particular parts of the laboratory work had presented learning difficulties. Second, taped interviews were conducted in the middle and at the end of the course, with five volunteer individuals assigned to probe issues in some depth. The five students that we eventually chose for the interviews were drawn from a list of eleven volunteers. In choosing the students we wished to have a mix of gender (two males and three females), a mix of attitudes to laboratory work (one was terrified of it, two did not want to be doing it, one was prepared to enjoy it, and one was looking forward to it), and we wanted one mature-age student.

Findings

Presage

The most telling part of our first class survey was the high percentage (40 percent) of students who were unable to muster any positive feelings toward laboratory work. Even before entering the door of the laboratory, these students would be at a learning disadvantage. Some of the comments were particularly disturbing because of the strongly emotive negative attitudes expressed. The following extracts provide illustrative examples.

- "Something always seems to go wrong *and* I don't sleep well the night before."
- "I hate lab work—just a necessity of my course that I have to do."

Some 36 percent of respondents were looking forward to the course for a variety of reasons that spanned areas such as

- personal enjoyment of hands-on work (e.g., "I prefer practical work to straight theory"),
- pedagogical pragmatism (e.g., "A change from monotonous lectures"),

- career advantages (e.g., "Lab work leads to experience and knowledge, which may help in future employment"), and
- epistemological values (e.g., "Bigger-picture understanding of why we do the theory; I will learn new things").

A staggering 70 percent of the students not looking forward to laboratory work were also not enthusiastic about designing and carrying out their own experiments. This was in stark contrast to the students who were looking forward to laboratory work, 86 percent of whom were also looking forward to some independence in the laboratory, giving reasons such as the following:

- "Following a lab manual is good (and easy), but seeing if something works in my own design is exciting and challenging (also frustrating)."
- "Should be a great learning experience, but also quite daunting; means we *have* to have a good understanding."
- "Doing real scientific work is more appealing than following steps with no idea of what's going on."
- "I like to problem-solve and think for myself."

These latter responses reflect higher-order attitudes toward learning and a desire of the students to take control of their own learning. Thus students with good prior experiences seem to enjoy laboratory work and show a willingness to embrace laboratory sessions that move them away from teacher-centered learning and provide them with opportunities for some independent thought. The kinds of experiences that our students nominated as leading to good feelings about laboratory work included not being made to feel inadequate by teaching staff after making mistakes or not understanding something, not having to repeat work over and over to get a "correct" result, having time to reflect on what was being done and why, feeling confident with basic procedures (e.g., how to use a pipette, how to assemble a gel rig), and not being bored with the work.

Process

Perceived Relevance. A factor that appeared to stimulate interest in the course, and hence provide an environment in which students wanted to take charge of their own learning, was the perceived relevance of what they were doing. Comments such as "Lab subjects give students an idea of a working biology lab environment" and "Lab work is important to most professions" indicated a holistic conception of the place of laboratory work in science but were not commonly expressed. Unfortunately, students often do not share the belief with their teachers that laboratory work is useful to their professional or even everyday life (Wilkinson and Ward, 1997). This perception may be linked to student problems in understanding the laboratory work or to the fact that it actually is not useful (i.e., poor judgment by the teacher in selecting the particular laboratory work). We found, especially, that students who had already decided on their career paths could be unwilling learners if the laboratory work was not directly of use to those paths. Pedagogically, this problem is hard to overcome since it requires an almost individual approach to help the students to find their own relevance and to think more holistically.

The top four things that the students were most hoping to learn from the course were laboratory techniques (85.3 percent), experimental procedures (85.3 percent), use of specialized equipment (80.4 percent), and understanding concepts already studied (73.1 percent). By the end of the course, the top four things that the students claimed to have learned were laboratory techniques (88.6 percent), understanding concepts already studied (80 percent), use of specialized equipment (74.3 percent), and experimental design (74.3 percent). Thus, very broadly, there was congruence between expectation and actual outcome of major learning issues.

How Students Learn. An important question that we asked the cohort of five students was what their idea of learning was. All gave answers that reflect a conception of at least some level of deep approach to learning, and their responses (see Table 3.1) fit into the six conceptions outlined by Prosser and Trigwell (1999). Sadly, however, all of the students expressed the view that learning at the university, especially in the early years, is very much based on memorizing facts.

With regard to how the students thought they learned practical work, Table 3.2 indicates that the major pathway was via asking someone perceived to be knowledgeable. What our survey failed to probe was whether the students simply sought a correct answer to a problem, asked for an explanation, or, better still, were guided to a better understanding by, for example, constructivist techniques. The observations of Raghubir (1979) are still as relevant now as they were

TABLE 3.1. Interview Responses to What "Learning" Means

Categories of conceptions of learning	Student responses
Quantitative increase in knowledge	
Memorizing	
Acquisition for subsequent utilization	"I guess it's acquiring the knowledge and the ability to apply knowledge to a situation."
The abstraction of meaning	"Failing is a very important part of learning and that can probably be related to lab work so when something goes wrong it kind of pushes everything back down in your brain and you get to think about it more and it sticks more. . . . I don't think learning means just go away and learn this list of things . . . it means having an aim underneath, understanding the concept and being able to relate it to something."
Interpretive process aimed at understanding reality	"Learning to me is to be able to understand something enough so that you can see the big picture."
Developing as a person	"It stretches your mind. . . . If you learn something, you are not the same person you were before."
	"I think it's really about personal development rather than vocational knowledge."

Note: The responses have been grouped according to the categories outlined by Prosser and Trigwell (1999).

in 1979. He noted that the failure of laboratory students to meaningfully learn and retain information had much to do with the attitudes of well-meaning but misguided academic staff:

> Teachers tell students too much; they deprive them of the opportunity to learn for themselves. In the laboratory, for example, they are likely to tell them just about everything—how to assemble an apparatus, how to design an experiment, and what outcomes to expect. Of course they think they do this for a good reason—to save time and to save the experiment.

TABLE 3.2. Responses to the Question, "How Do You Learn Practical Work?" (*n* = 35)

Number of responses	Percentage of total	Choice of answers
33	94.3	Asking teaching staff questions
31	88.5	Reading the laboratory manual
28	80.0	By understanding what I am doing
27	77.1	Talking it over with partner/other colleagues
23	65.7	Watching others (either demonstrators or other students)
20	57.1	Going over troubleshooting tips
19	54.2	By thinking through the principles
17	48.6	Repeating procedures until I am confident, i.e., practice
15	42.6	Going back over own notes from lectures and labs
11	31.4	Attending lab revision sessions
8	22.8	Reading textbooks
8	22.8	Doing examples of calculations
7	20.0	Going over past exam papers
3	8.5	Memorizing facts

Note: Students were asked to check off as many appropriate answers as they liked.

Hodson (1995) further observes that if students do not have the right understanding to make appropriate observations or interpretations, it is actually damaging for teachers to provide solutions too readily. The problem may be that the students have a different understanding, and merely telling them they are wrong will not help them to alter their misconceptions. Although reading the laboratory manual was also nominated as a major learning pathway, only 54 percent of the students claimed to have read the relevant part of the manual 80 to 100 percent of the time prior to a laboratory class. It would seem that many students enter a laboratory session without a clear idea of what they are going to be doing or why, and this places them at a serious learning disadvantage.

It is accepted that the active, deep learning fostered by a more open-ended approach to laboratory work is enhanced by group interaction (Stefani and Tariq, 1996; Fraser and Deane, 1998; Friedman, 1999; Lawson, 1999). The students in our class worked in pairs and results from the surveys and interviews showed that this was valued by the students. Before laboratory sessions, conscientious pairs would get together to plan what needed to be done and discuss any problems in understanding concepts. During all laboratory sessions, the lively and relevant discussions within pairs, as well as a willingness to share information between pairs, was testament to the positive influence of group work on student-centered learning. Having a partner to work with also cut down considerably on the mundane questions (e.g., Which pipette do I use?) that might otherwise have been put to the teaching staff. However, while having a partner is useful, it became clear to us that this system was open to an abuse that actually discouraged learning. Some pairs worked solely on the principle that finishing early was the most important aspect of each class, and they would often divide up the work so that each would do only half of the work—presumably critical results were shared at some later stage. Although this attitude represents the ultimate in cooperative sharing of the workload, it has its shortcomings in the practical assessment component when some students were faced with trying to decipher, set up, or troubleshoot an experiment they had never seen before because their partners had been the ones who dealt with it during the one and only laboratory session when it was presented. After the exam, one student of a pair who had taken this approach actually commented that it "shows how doing it yourself does really help with learning."

One of the questions on our second survey asked students to nominate which areas of the course presented some learning difficulties and how they tried to overcome the problems. We provided a list of twenty-eight aspects of the course that covered all of the laboratory work done, from which they could choose as many as they liked. The four areas presenting the most learning difficulties are given in Table 3.3, together with the ways in which the students attempted to overcome the difficulties. These indicate that learning difficulties were perceived more with conceptual aspects of the laboratory work than with technical or computational aspects. In trying to overcome the learning difficulties, the students used some of the techniques em-

TABLE 3.3. The Four Most Prevalent Learning Difficulties Experienced by Students in Lab Work ($n = 35$)

No. of respondents	Type of problem	Method used to overcome problem
13 (37.1%)	Designing a competitive ELISA assay	• Practicing • Drawing it out to understand what was happening at each step • Talking to classmates • Going through troubleshooting tips; designing another protocol • Trying again • Asking demonstrator • Thinking more • Reading and learning ELISA concepts
11 (31.4%)	The concept of a competitive ELISA assay	• Perseverance, practice • Talking to classmates • Actually doing it in the lab • Reading about it in the lab • Talking to demonstrator • Undertaking ELISAs • Learning it again • Reading lab manual
10 (28.5%)	The concept of titre	• Practicing • Reading lab manual • Readings • Asking demonstrator to go through it on the board
9 (25.7%)	Interpreting different kinds of electrophoresis gels	• Asking demonstrator • Practicing and perseverance • Going to the lab revision session • Collaborating with others

ployed for general learning (see Table 3.2), i.e., asking the demonstrators to explain, reading, and talking to classmates. However, another learning approach was also adopted and encompassed the broad area of repetition—practicing and reviewing. One student from the cohort of five actually stated emphatically, "I think the only way you really can learn (laboratory work) is practice." This approach is part of the concept of creating time for reflection and reconciling theory and practice, which is regarded as essential in generating pedagogically useful laboratory work (e.g., Hegarty-Hazel, 1990a; Hodson, 1995).

An insight into the importance of laboratory work in creating episodes (White, 1991) that enhance the learning experience through clarifying concepts was provided by the laboratory sessions devoted to the ELISA technique. ELISA is an immunological method for the detection of a wide range of human, livestock, and plant diseases, especially those that are viral in nature (e.g., Newcastle disease, a highly contagious infection that gives chickens flulike symptoms; stock suspected of being infected has to be destroyed, as occurred in isolated outbreaks in Australia in 1998). The last question at the end of both interviews with the cohort of five students asked a very specific learning question about the ELISA technique. In the second interview the students were also asked where they had drawn the information from to answer the question. It was interesting that their answers at both interviews indicated a strong reliance on theory gained from past lectures, but there was also a demonstrable reliance on the laboratory work to reinforce the theory and to learn from doing something rather than reading or hearing about it. One student commented on the practical work in which students had to design, optimize, and interpret a competitive ELISA for the detection of Newcastle disease virus in four chicken samples:

> I thought the whole practical was wonderful for a number of reasons, ranging from the fact that we had an opportunity to actually design a lab for ourselves . . . to the everyday significance of it (I still have strong memories of the television scenes in 1998 where thousands of chickens had to be destroyed because of the virus) to the delightful names that the lab staff gave our chickens. . . . I will probably never forget the principles that we used for the competitive ELISA.

Product

The assessment for the course was, unfortunately, constrained within the limits imposed by a higher level of departmental organization and required a summative, classificatory approach. However, the practical exam (total duration, sixty-nine minutes) was devised such that it largely tested understanding and focused on deep learning and application. It was structured such that there were fifteen stations set up in the laboratory, at which each student could spend three minutes to write an answer to the question(s). The following example of a question is indicative of the way in which the exam was structured to test understanding: "Is this apparatus set up correctly for performance of the transfer step of a Western blot? Comment." There also was a single twelve-minute station at which the students had to perform a procedure involving preparation and reading of spectrophotometer samples, and there were four three-minute stations at which there were no questions and the students had time to reflect on previous questions.

The students reported that they undertook various strategies to study for this exam, ranging from hardly doing anything and just relying on knowledge assimilated during class time to undertaking a very detailed revision of laboratory notes and likely questions, e.g.,

> I read through the procedures and things and also I studied quite a bit on ELISAs, making sure that I understood the principles of them. I read through the troubleshooting section on ELISAs. . . . I had a written list of what the question areas were that were put up on the notice board and I suppose I just read up as much as I could on what those areas were and what we did.

Although most of the students felt reasonably confident going into the practical exam, for some others in the class the examination process loomed as something of a nightmare. One student commented to us, "There were a couple of people having absolute heart attacks on the day of the exam in the common room, just absolutely climbing the walls." This rather colorful description is exaggerated, but it does serve as a reminder to academic staff that some students approach examinations with considerable trepidation. The exam marks were generally pleasing, a factor probably influenced by a practice exam, held the week before, which gave a familiarity with the exam layout, instilled a notion of the time constraints, and provided good examples of question format, such that the students were better mentally prepared. The time constraints imposed on each question were, undoubtedly, a significant cause of anxiety but, otherwise, most students thought the practical exam was a fair test of the laboratory work covered during the semester. The reason for the time constraints stemmed from both the logistical aspects of setting up and running a comprehensive practical exam for sixty-six students and the philosophy of the course convenor, who believed that students need to appreciate the necessity for expediency in analysis and diagnostic accuracy. Although it is a valid point that students need to learn how to work under pressure, this learning experience, ideally, should not be introduced at the point of examination.

CONCLUSION

It cannot be denied that laboratories present a unique and valuable hands-on approach to learning science. However, the wealth of literature, along with our own study, indicates that the poor student perception of laboratory work may often be lowering the educational value of this tool. It is thus important to engender attitudes that are conducive to learning. The literature lists many conditions that promote or inhibit interest and satisfaction in laboratory work (see, e.g., Gardner and Gauld, 1990), including laboratory facilities (quality of resources), time allowed, variety, cognitive challenge, integration of theory with laboratory work, teachers' abilities, autonomy, and social interaction. Our study, using the Prosser and Trigwell (1999) presage-process-product model as a framework, has concurred with these conclusions and also provided some insight into what students bring to the laboratory and how this influences their learning. Our main findings were as follows:

- Bad or unpleasant past experiences in the laboratory can create attitudes that are not conducive to learning. Students need to feel unpressured and comfortable in their laboratory surroundings if they are to maximize their receptivity to learning. Students should never approach laboratory work with fear or hatred, yet this can occur. Our results strongly indicate that a positive contribution to attitudes toward laboratory work can be achieved by having patient, caring, and empathetic staff; close interaction with a partner during laboratory sessions; and time during sessions for the students to think about and reflect on what they are doing and what they are learning.

- Our interviews with a cohort of five students showed that, even in those students possessing a conception of a deep approach to learning, shallow learning is recognized as an expedient way of passing tertiary assessment. The problem of perceived relevance, along with the pressures of time, other subjects, and other commitments, take their toll on all but the most genuinely engrossed students. However, laboratory course designers can encourage deeper learning by having more experiments in which the students are given some independence in designing protocols and encouraging students to be critical, to raise their own questions and offer alternative perspectives. A particular emphasis in our laboratory sessions was troubleshooting, where the students were encouraged to learn for themselves how a particular technique could go wrong and how an experimental situation could be rescued; mistakes in calculations, techniques, setting up equipment, or designing experiments were used as a pedagogical tool to promote discussion in a nonthreatening environment. In this type of learning environment it is actually possible to transform student attitudes from negative to positive. One student at the end of the course was moved to comment,

> I liked this lab (I normally dislike them) because it focused on understanding and learning of the theory behind the laboratory and also the practical component. . . . I have been converted from a lab hater to a lab liker just because of the different approach taken in this lab subject.

- Working in pairs or groups stimulates open discussion. If you can be convinced by your partner's argument that an ELISA plate needs to be set up in a particular way to answer a particular question posed in the practical work, then both your partner and you have engaged in the valuable scientific inquiry processes of logical and critical thinking.
- Our study also showed that, although asking questions of someone knowledgeable was the key learning strategy, the more motivated students used practice and repetition as a valuable learning pathway. For the course designer, this indicates that time should be set aside (either during or between laboratory sessions) for students to be able to take advantage of this. At least some laboratory experiences should also be conceived in

terms of activities that form useful episodes (White, 1991) for students so that there is opportunity for them to link these with theory-centered applications.

The relevance of the quotation at the beginning of this chapter is widely accepted among science teachers and behooves us to use the laboratory as an important pedagogical tool for nurturing scientific inquiry. This can be done only in an environment that encourages and stimulates learning.

REFERENCES

Anderson, O.R. (1976). *The experience of science: A new perspective for laboratory teaching.* New York: Teachers College Press.

Australian Education Council (1994). *Science: A curriculum profile for Australian schools.* Melbourne, Australia: Curriculum Corporation.

Bain, J. (1994). Understanding by learning or learning by understanding: How shall we teach? An inaugural professorial lecture by the professor of teaching and learning, Griffith University, Faculty of Education, Brisbane, Queensland, Australia. September.

Baird, J.R. (1990). Metacognition, purposeful enquiry and conceptual change. In E. Hegarty-Hazel (Ed.), *The student laboratory and the science curriculum* (pp. 183-200). London: Routledge.

Barrie, S.C., Buntine, M.A., Jamie, I., and Kable, S.H. (2001). APCELL: Developing better ways of teaching in the laboratory. In *Proceedings of research and development into university science teaching and learning: Annual workshop* (pp. 23-28). Sydney, Australia: University of Sydney, UniServe Science.

Boud, D. (1986). Aims, objectives, and planning. In D. Boud, J. Dunn, and E. Hegarty-Hazel (Eds.), *Teaching in laboratories* (pp. 13-34). The Society for Research into Higher Education and NFER-Nelson: Guildford, United Kingdom.

Feteris, S., Gunstone, R., and Mills, D. (2001). Learning in the laboratory—Research on student and staff perceptions [abstract only]. In *Proceedings of research and development into university science teaching and learning: Annual workshop* (p. 47). Sydney, Australia: University of Sydney, UniServe Science.

Fraser, S. and Deane, E. (1998). *Doers and thinkers: An investigation of the use of open-learning strategies to develop life-long learning competencies in undergraduate science students.* Canberra, Australia: Department of Employment, Education, Training, and Youth Affairs. Evaluations and Investigations Programme, Higher Education Division, Commonwealth of Australia. Australian Government Publishing Service.

Friedman, D.L. (1999). Science, YES!—A program to excite the scientist in the teacher. *Journal of College Science Teaching,* 28: 239-244.

Gardner, P. and Gauld, C. (1990). Lab work and students' attitudes. In E. Hegarty-Hazel (Ed.), *The student laboratory and the science curriculum* (pp. 132-156). London: Routledge.

Gunstone, R. (1991). Reconstructing theory from practical experience. In B.E. Woolnough (Ed.), *Practical science: The role and reality of practical work in school science* (pp. 67-77). Milton Keynes, PA: Open University Press.

Hegarty-Hazel, E. (1986). *Lab work. SET: Research information for teachers #1.* Canberra: Australian Council for Education Research.

Hegarty-Hazel, E. (1990a). Life in science laboratory classrooms at tertiary level. In E. Hegarty-Hazel (Ed.), *The student laboratory and the science curriculum* (pp. 357-382). London: Routledge.

Hegarty-Hazel, E. (1990b). The student laboratory and the science curriculum: An overview. In E. Hegarty-Hazel (Ed.), *The student laboratory and the science curriculum* (pp. 3-26). London: Routledge.

Hodson, D. (1995). Toward a more critical approach to laboratory work. Paper given at Conference for the Australian Science Teachers Association, CONASTA 44. University of Queensland, Brisbane, September 24-29, 1995.

Hodson, D. (1996). Laboratory work as scientific method: Three decades of confusion and distortion. *Journal of Curriculum Studies, 28:* 115-135.

Lawson, A.E. (1999). What should students learn about the nature of science and how should we teach it? *Journal of College Science Teaching, 28:* 401-411.

Maienschein, J. with students (1998). Scientific literacy. *Science, 281:* 917.

Prosser, M. and Trigwell, K. (1999). *Understanding learning and teaching.* Buckingham, England: The Society for Research into Higher Education and Open University Press.

Raghubir, K.P. (1979). The laboratory investigative approach to science instruction. *Journal of Research in Science Teaching, 16:* 13-17.

Schwab, J.J. (1962). The teaching of science as enquiry. In J.J. Schwab and P.F. Brandwein (Eds.), *The teaching of science* (pp. 1-103). Cambridge, MA: Harvard University Press.

Snyder, B. (1971). *The hidden curriculum.* New York: Knopf.

Solomon, J. (1994). The laboratory comes of age. In R. Levinson (Ed.), *Teaching science* (pp. 7-21). London and New York: Routledge.

Staer, H., Goodrum, D., and Hackling, M. (1998). High school laboratory work in Western Australia: Openness to inquiry. *Research in Science Education, 28:* 219-228.

Stefani, L.A.J. and Tariq, V.N. (1996). Running group practical projects for first-year undergraduate students. *Journal of Biological Education, 30:* 36-45.

White, R.T. (1991). Episodes, and the purpose and conduct of practical work. In B.E. Woolnough (Ed.), *Practical science* (pp. 78-86). Milton Keynes; PA: Open University Press.

Wilkinson, J. and Ward, M. (1997). A comparative study of students' and their teachers' perceptions of laboratory work in secondary schools. *Research in Science Education, 27:* 599-610.

Zadnik, M. and Yeo, S. (2001). Improving teaching and learning in undergraduate science: Some research and practice. In *Proceedings of research and development into university science teaching and learning: Annual workshop* (pp. 12-16). Sydney, Australia: University of Sydney, UniServe Science.

Chapter 4

Developing the Metacognitive and Problem-Solving Skills of Science Students in Higher Education

Rowan W. Hollingworth
Catherine McLoughlin

INTRODUCTION

In tertiary education nowadays a greater emphasis is being directed toward the development of generic skills or graduate attributes, including communication skills, global perspectives, problem solving, teamwork, and social responsibility. Reliance on a content-based curriculum in science is not an appropriate preparation for the rapidly changing world of the future (Lowe, 1999). The new emphasis on generic skills is aimed at addressing an urgent need for professionals who can find realistic solutions to complex, real-world problems. Jonassen (2002) goes so far as to state, "I believe that the only legitimate goal of professional education, either in universities or professional training, is problem solving" (p. 78). It is clear, then, that tertiary educators need to carefully examine their methods of teaching problem-solving, as well as the types of problems they select for their students, if they wish to produce graduates effective in the modern workplace, society, and life in general.

In this chapter the development of problem-solving skills, specifically in the context of first-year university-level science subjects, is discussed. This is done in the context of the broader profile of students now entering universities and studying in both the on-campus and off-campus (distance-education) modes. The types of problems which students may tackle in their learning activities are considered. The teaching/learning environments in which learning tasks are carried out and which will effectively support the development of good problem-solving skills are examined. Finally, a number of approaches to developing the higher-order cognitive skills of first-year science students, including an online tutorial, metAHEAD, are described.

LINKING PROBLEM-SOLVING AND METACOGNITIVE SKILLS

Broadly speaking, the development of effective problem-solving skills depends on two factors:

- The teaching/learning environment in which the problem-solving skills are developed
- The types of problem students are exposed to

In terms of pedagogy, it is not sufficient to rely on a "transmission of knowledge" approach, which relies heavily on regurgitation of facts and concepts and the solution of routine problems or exercises. This type of approach targets only lower-order cognitive skills (Zoller, 2000). Even so, the traditional "show and tell, then practice" approach to problem solving may help students develop some fluency at handling and applying concepts. However, this simplistic approach can no longer be relied on as the sole strategy or even a very effective strategy for building the higher-order cognitive skills required to tackle the more ill-defined, complex, interdisciplinary problems university graduates now face when they enter the workforce.

Recent analysis of research on the teaching of problem solving has shown that knowledge of strategy and practice of problem solving has had little effect on student performance and achievement, whereas effective approaches to teaching problem solving all gave attention to contextualized strategies related to the knowledge base. The learning conditions recognized as significant for building problem-solving skills are those that provide learners with guidelines and criteria they can use in judging their own problem-solving processes and products (Everson and Tobias, 1998). A repertoire of learning strategies, a capacity to manage one's own learning, and an awareness of one's own knowledge and skills are fundamental in order to learn effectively and to problem solve in a variety of contexts. The range of skills relating to the self-management of learning as students engage in monitoring and evaluating their own problem solving is known as *metacognition.*

Development of Metacognition

Metacognition involves both knowledge of cognition (the learners' knowledge about their own processes of cognition) and regulation of cognition (the ability to monitor and control those processes) (Metcalfe and Shimamura, 1994; Schraw, 1998). Metacognition can be viewed as a supervisory or metalevel system that controls and receives feedback from normal information processing. Metacognitive *knowledge* refers to what the learner knows and understands about the task in hand, while metacognitive *regulation* refers to the strategies the learner uses to complete the task. This regulation involves planning, organizing, and monitoring the task, but it also involves evaluating outcomes and reflecting on learning and problem solving.

The literature attests to the fact that even at beginning tertiary level, few students appear to have developed the expert problem-solving and metacognitive skills that enable them to cope effectively with learning independently and successfully in the sciences (Volet, McGill, and Pears, 1995; Everson and Tobias, 1998; Gourgey, 1998). There is evidence that metacognitive skills can be taught, although the range of programs and approaches attempted has been varied. Gredler (1997) proposes three essential conditions, which apply for training and development of metacognitive skills:

- The training should involve students' awareness of what the process involves, as this makes them participants in the process.
- The performance criteria used for evaluation of achievement should match the kinds of metacognitive activities addressed in the instruction.
- Metacognitive training should provide support for engagement in metacognitive activities.

Masui and De Corte (1999) propose similar conditions, suggesting an integrated set of instructional principles for an effective learning environment to enhance metacognition and problem-solving skills for university students:

- Embed acquisition of knowledge and skills in a real study context.
- Take into account the study orientation of students and their need to experience the relevance of the learning and study tasks offered to them.
- Sequence teaching methods and learning tasks and interrelate them.
- Use a variety of forms of organization or social interaction.
- Take into account informal prior knowledge and individual differences between students.
- Learning and thinking processes should be verbalized and reflected upon.

Examples of studies of metacognitive development at the university level are rather scarce. They typically have involved lecturers and instructors in long-term, face-to-face situations over a period of at least one semester. These include a study of first-year computer science students' development of a metacognitive strategy and coaching its use in a socially supportive environment (Volet, 1991), and a self-directed learning program to develop transferable learning and metacognitive skills for first-year chemistry students (Zeegers, Martin, and Martin, 1998). Outside the science area, Masui and De Corte (1999) have examined the trainability and effect on academic performance of enhancing the learning and problem-solving skills of business economics students.

There is consensus in the research literature that learners have an opportunity to evaluate the outcome of their efforts, to reflect on and self-assess their own approaches to learning. Simply providing knowledge without experience or vice versa does not seem to be sufficient for the development of metacognitive control. The most effective metacognitive instruction schemes in the literature involve providing the learner with knowledge of cognitive processes and strategies, together with experience or practice in using them (Boekaerts, Pintrich, and Zeidner, 2000).

TEACHING PROBLEM SOLVING

The focus in this chapter is directed more to the field of science education and the rapidly developing field of educational technology, rather than the field of cognitive psychology, in which human problem solving is a well-established and wide research area. Much of the research on the training of problem solving and metacognition has involved primary- and secondary-level learners, rather than those at the tertiary level.

Sadly, it appears that this large research resource has been only lightly tapped by the majority of university science teachers. Gabel (1994) provides thorough reviews of research on problem solving in several science subjects, which had been carried out up to the mid-1990s. More recently Taconis, Ferguson-Hessler, and Broekkamp (2001) analyzed articles researching the effectiveness of teaching strategies for science problem solving as reported in twenty-five high-quality international journals between 1985 and 1995. From approximately 2,000 papers in these journals, forty experiments in twenty-two papers on the teaching of problem solving were used. These studies concentrated on the cognitive aspects of the teaching interventions and left the metacognitive aspects implicit.

The traditional approach to problem solving in science has been to engage students in repetition of routine exercises, with an emphasis on the use of algorithms and sequences of steps rather than strategies and reflection on processes. Hobden (1998) suggests this has been used uncritically as a teaching strategy "on the optimistic assumption that success with numerical problems breeds an implicit conceptual understanding of science" (p. 219). This approach may help students develop routine expertise, that is, speed and accuracy at routine problem solving, but will fail to develop adaptive expertise, the ability to reflect on strategies or to adapt to solving new problems in a flexible manner (Hatano and Inagaki, 1986).

From their analysis, Taconis, Ferguson-Hessler, and Broekkamp (2001) found that attention to the structure and function of the knowledge base was a feature of effective metacognitive training, while attention to knowledge of strategy and the practice of problem solving appeared to have little effect. The learning conditions recognized as significant for building problem-solving skills are those which provide learners with guidelines and criteria they can use in judging their own problem-solving processes and products. The provision of immediate feedback to learners is also essential. These conclusions are congruent with earlier research carried out by researchers in the field of metacognition (Alexander and Judy, 1988; Clarke, 1992; Lajoie, 1993).

THE NEED FOR ILL-DEFINED PROBLEM TYPES

The tendency toward repetitive practice at routine problem solving as the traditional approach in science teaching has been highlighted previously. The need for science graduates to be able to solve the complex, interdisciplinary, real-world problems they will face when they enter the workforce was also alluded to by Zoller (2000). The types of problems students are asked to solve must be a major consideration in developing transferable problem-solving skills.

To this end, Jonassen (2000) has provided a classification of problem types and their characteristics. The problem types identified include logical, algorithmic, rule-using, decision-making, troubleshooting, diagnosis, case analysis, design, dilemmas, and stories. For the purposes here it is sufficient to consider a range of problems varying along a continuum from well-defined to ill-defined. Well-defined problems are those that are typically seen at the end of chapters in student textbooks, and are designed for self-study and reinforcement of key concepts. Usually they present all the elements of the situation; demand a limited number of skills, rules, and principles; and require a correct solution through a prescribed solution process in a well-defined domain of knowledge. In contrast, ill-defined problems require students to interpret some of the problem elements and may possess multiple solutions or approaches. As it may be unclear which rules or principles are necessary for a solution, the learner needs to think strategically, employ metacognitive skills, and defend his or her solution. In general, any particular problem would fall somewhere on a spectrum ranging between the two extreme problem types. Table 4.1 indicates some differences in the characteristics of well- and ill-defined problems. The development of high-level problem-solving skills requires that students be given open-ended or ill-defined problems.

In the specific context of calculation-type chemistry problems, Johnstone (1998) has classified problems along three dimensions: whether data are complete or not, whether the method is familiar or not, whether the solution or goals are given or open. This classification results in eight types of problems, increasing in difficulty as data become incomplete, the method becomes unfamiliar, and the solution

TABLE 4.1. Some Characteristics of Different Types of Problems

Characteristic	Well-defined problem	Ill-defined problem
Data	Complete	Incomplete or not given
Knowledge domain	Well-defined	Ill-defined
Rules and principles	Limited rules and principles in organized arrangement	Uncertainty about concepts and principles necessary for solution
Solution process	Familiar; knowable, comprehensible method	Unfamiliar; no explicit means for action
Answer	Clear goal, convergent; possess a correct answer	Uncertain, multiple or no solution; need to make judgments and evaluations

becomes open. The advantages to the students in solving each type of problem are also outlined, with higher-order cognitive skills being developed as more ill-defined problems are tackled.

Some specific examples of first-year chemistry problems follow to illustrate a range of problem types.

Problem 1: Excited hydrogen atoms produce many spectral lines. One series of lines, called the Pfund series, occurs in the infrared region. It results when an electron changes from higher levels to a level with $n = 5$. Calculate the wavelength of the lowest energy line of this series.

Problem 2: Exposure to high doses of microwave radiation can cause damage. Estimate how many photons with $\lambda = 12$ cm must be absorbed to raise the temperature of your eye by 3.0°C. How long would it take for your eye to be heated this much if it was placed near a typical but defective microwave oven in such a position that it received one-hundredth of the power output of the oven?

Problem 3: We view visible light with our eyes every day without any ill effects. However, exposing our unprotected skin, which is much less "sensitive" than our eyes, to a day on the beach can have very painful consequences. What is happening in each of these circumstances at the molecular level? Give as many details of the chemistry as possible and include quantitative calculations, if possible, to back up the explanations.

Problem 1 is a typical end-of-chapter exercise, with one correct answer, which can be obtained through application of an algorithm. All of the data (except for values of fundamental constants) required for the calculation are supplied. The problem has little relation to any phenomena a student can easily relate to. To answer Problem 2, students will need to determine what extra data are required for a solution and then search to find this information. (Data on the specific heat capacity of the eye and the power level of a microwave oven are needed and might have to be estimated by some sensible approximation.) Reasonable values for the answers are required, rather than exact numbers. The setting of the problem relates more to student experience than Problem 1. Problem 3 also relates to students' everyday experience. The way students answer the question depends much more on what they decide to do and research. It is an open-ended question and clearly no one answer is correct. This question has been used as a collaborative question, where students work as a team to produce their answer.

It is not suggested that practicing routine, well-defined problems does not benefit student learning. Recall that the analysis of Taconis, Ferguson-Hessler, and Broekkamp (2001) showed that well-structured domain knowledge is important in problem solving. Practice in solving well-defined problems, in a properly supportive learning environment, can help this development. At first-year level, when students are still struggling to build a well-organized knowledge base in their science subjects, it is important to allow students sufficient time and practice with concepts in order to develop meaningful understanding. It is all too easy to burden students with more and more content, facts, and inert knowledge. Mayer (1997) suggests that it is a mistake to believe in the idea of *prior automatization,* namely, that students can only develop higher-order thinking skills after they have mastered the prerequisite lower-order skills. Hence it is imperative that students are given the opportunity to tackle real-world problems and not simple routine exercises. There is evidence that further skills are required for success in ill-defined problem solving. Shin, Jonassen, and McGee (2003) have found that domain knowledge and reasoning skills are significant predictors of well-defined problem-solving scores, whereas regulation of cognition and attitudes toward science are additional significant predictors of problem-solving success for ill-defined problems.

DESIGN OF TECHNOLOGY-SUPPORTED
METACOGNITIVE TRAINING

How can problem-solving skills and instructional design principles develop metacognitive awareness be implemented in a technological environment? Jonassen (1997) has proposed different instructional design models for learning well-defined and ill-defined problem solving. For well-defined problems, six steps are suggested:

1. Review prerequisite concepts, rules, and principles.
2. Present a conceptual model of the problem domain.
3. Model problem-solving performance with worked examples.
4. Present practice problems.
5. Support the search for solutions.
6. Reflect on the problem and the solution.

Different steps are suggested for ill-defined problems:

1. Articulate the problem context.
2. Introduce problem constraints.
3. Locate, select, and develop cases for learners.
4. Support knowledge-base construction.
5. Support argument construction.
6. Assess problem solutions.

Considering more specifically the development of higher-order cognitive skills, metacognition, and reflection, three important implications of social constructivist theory should be built in to the learning environment from the beginning. First, in order to foster reflective thinking, students need *multiple sources of feedback* on their understanding gained through social interactions. Second, reflective thinking will most likely occur in situations where *problems are complex* and meaningful to the student. Third, *reflective thinking* requires students to organize, monitor, and evaluate their thinking and learning to come to a deeper understanding of their own processes of learning (Elen and Lowyck, 1999). In accordance with these principles, Lin and colleagues (1999) prescribe four features that can provide scaffolds to enhance reflection in the technology-based environment:

- *Process displays:* where students are explicitly shown what they are doing in performing a task
- *Process prompts:* where students are asked to explain what they are doing at different stages throughout their problem-solving procedure
- *Process modeling:* where students have access to databases and audio or video displays explaining what, how, and why other students and experts do what they do in solving a specific problem
- *Reflective social discourse*: where students share their learning experiences and gain feedback from a community of learners, for example, through an online discussion space

Instructional designers and teachers need to be able to adapt these suggestions and principles to the particular learning environment being built, based on their own expertise and judgment in what is in essence the solution of an ill-defined problem.

There are several examples of technology-supported approaches to metacognitive skills and problem solving. LUCID (Learning and Understanding through Computer-based Interactive Discovery) is a new model for computer-assisted learning workshops to promote student engagement in the learning process (Wolfskill and Hanson, 2001). LUCID provides students with an orientation for the learning process and allows them to freely navigate through activities. Key questions are employed to guide the exploration of interactive models and the development of understanding, with instant multilevel feedback (for questions with a single correct answer) to promote confidence while developing problem-solving skills. More complex questions network reporting and peer-assessment, promoting critical review and the achievement of consensus in a group. To assist in reflection and self-assessment, performance distributions are provided on the quantity and quality of the work and reports of other teams. This example illustrates how information and computer technologies (ICT) can enhance learning activities and the importance of using sound pedagogical principles to drive the implementation of ICT.

In a study conducted by Lin and Lehman (1999), biology students worked in a computer-based biology simulation learning environment, designing and conducting experiments involving the control of variables. The study investigated the use of explicit prompts to engage students in metacognitive thinking and problem solving involving control of variables. Students were given different types of prompts: reasoned justification, rule-based, or emotion-focused. Qualitative data showed that the reasoned justification prompts directed students' attention to understanding when, why, and how to employ experimental design principles and strategies, and this in turn helped students to transfer their understanding to a novel problem (Lin and Lehman, 1999).

METAHEAD: AN ONLINE TUTORIAL TO SUPPORT METACOGNITION

The changed profile of students currently entering universities in Australia is creating pressure to change teaching practices. Many students are mature age and work part-time while studying, creating a demand for more courses and programs in the distance-education mode. Many students have been away from study for a number of years and may not have the well-developed learning skills required for tertiary study and may have a limited background in science. Moreover, distance students may feel a sense of isolation in not being able to work cooperatively with other students or to compare the quality of their own work in relation to that of others enrolled in their courses. This requires university teachers to support learning in what is a new area of education for many first-year students. The metAHEAD tutorial, an online tutorial provided for both on-campus and off-campus students taking first-year biology, biophysics, or chemistry at the University of New England (UNE), was developed in this context.

There has been debate about the relative benefits of general study skills programs versus explicit skills training within subject teaching. Research indicates that the teaching of problem solving is best learned within the subject domain, rather than as a separate, decontextualized subject (Mayer, 1997). Research suggests that students are less likely to be interested in skills-development programs unless they can see direct application to the work they are doing in their subjects at the time they are engaged in a tutorial. The aim then was to maximize the attractiveness of the tutorial by highlighting the commonality of skills across biology, chemistry, and biophysics and starting with activities very similar to students' assignment tasks for various topics of study.

The objectives, in designing the tutorial, were as follows:

- To support the development of the metacognitive skills and habits of reflection, which are essential to effective problem solving in the sciences
- To foster students' problem-solving skills in first-year science, utilizing the communicative and supportive features of a technology-based environment
- To apply constructivist instructional design principles that can contribute to the development of an online environment to foster metacognition

Based on the current research on metacognitive training, a scheme for the development of metacognitive skills for science students that involved eight phases was adopted. The environment for metacognitive development utilized Web-based learning activities to engage learners in actual problem solving and reflection on their own problem-resolution strategies.

Phase 1: The concept of metacognition is operationalized. For the problem in question, students need to become aware of the problem-solving processes involved. For example, this requires analysis of the question, planning a solution, and selection of strategies and self-monitoring skills that can be used.

Phase 2: This phase involves the design of the problem environment. For particular problems in a topic in biophysics, for example, examine the different ways in which an expert and a novice student might answer the problem.

Phase 3: The problem is presented to the student.

Phase 4: Student responses are monitored to decide if any intervention (Phase 5) is required.

Phase 5: This step presents students with a scenario or problem where they are assisted in the processes and procedures of problem solving and made aware of their own problem-solving strategies.

Phase 6: Successful students are presented with further problems in the topic area to check whether they have transferred the strategies learned during Phases 3 and 4. If they have not, training continues.

Phase 7: Students are given the opportunity to reflect on their problem solving.

Phase 8: This final step involves a refinement of the training to create design guidelines for a problem-solving environment in different subject areas (biology, biophysics, and chemistry) in order to foster metacognition.

A flow chart diagram for these phases as applied in the metAHEAD tutorial is shown in Figure 4.1.

The overall structure of the tutorial comprises four modules, as shown in Figure 4.2. In the introductory module, Module 1, students encounter some basic ideas about learning and thinking, cognition, and metacognition. They also have the opportunity to take short questionnaires relating to problem solving and metacognition to help them assess their current level. Module 2 introduces students to con-

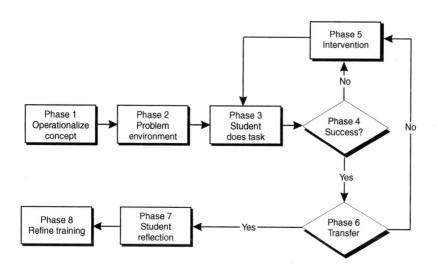

FIGURE 4.1. Phases for Metacognitive Training

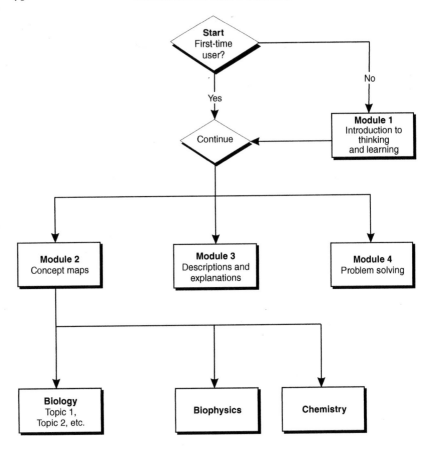

FIGURE 4.2. Modular Structure of metAHEAD Online Tutorial

cept mapping in each of the subject areas. In the other modules students start with the typical exercises they come across in the study of their subjects, leading to metacognitive skills development as they tackle further problems. These modules have similar parallel paths for each subject; Module 3 considers tasks involving explanations and descriptions, while Module 4 involves calculations-type problems. For a more detailed flow chart of the path through a module, see Hollingworth and McLoughlin (2001).

When students log on to the tutorial, they are asked to choose a subject and topic on which to work. They are asked to solve the problem, being presented with a variety of prompts and questions about the processes they engage in during their solution process. In this way, students can be exposed to a number of *process displays* and *prompts*. They have an opportunity to view answers given by other students, ranging from poor to very good answers. A model answer from the lecturer is also available. All these answers are commented upon and students may also listen to audio segments or video clips of other students and lecturers as they worked on the problems. In this way *process modeling* is provided.

For some questions, students are asked to collaborate with each other to build a communal answer on the bulletin board. There are opportunities for *reflective social discourse* through discussion of particular problems and also more general issues in bulletin board topics. Many of the same ideas are being applied, discussed, and reflected upon in different questions on different topics and subjects, which are believed to assist in transfer of the skills developed to broader areas of application. Throughout the tutorial students are prompted to answer questions and make notes in an online logbook in order to keep a record of their strategies and skill development.

Evaluation of metAHEAD

Oliver (1999) has remarked that while a range of methodologies for evaluation of ICT learning environments exists, each may be restricted in its use and in the range of situations to which it can be applied. An evaluation approach was sought that was broad and flexible enough to suit the situation, namely, the introduction of an innovative resource within a university context. The Open University model, as described by Jones, Tosunoglu, and Ross (1996), is a useful framework. This approach focuses on three main themes: context, interaction, and outcomes, as illustrated in Table 4.2. From the outset, data have been gathered on the design and use of metAHEAD from pilot studies, practitioners' opinions, instructional designers, and academic staff.

The gathering of a range of data for the evaluation of metacognition was facilitated by metAHEAD being Web-based. Data collected online included a self-rating quiz, taken at the beginning and at the end of semester; students' self-predictions of success with particular problems and reflection upon these on completion of the prob-

TABLE 4.2. Features of the Evaluation Approach Adopted

Feature	Context	Interactions	Outcomes
Rationale	Need for metAHEAD at UNE; curriculum context	Need to look at student interactions with the resource	Learning outcomes; problem-solving outcomes; changes of perception and attitude must be considered
Data	Designers aim; principles underpinning design; pressing iterations	Records of students' interactions, student diaries, and online logs	Measures of effective problem solving, changes in attitude, strategy, perception of self
Methods	Interviews; written records	Observations, videos, diaries, computer records, product data generated by students	Focus groups, tests, and questionnaires

lem; online logbook entries, including students' notes on strategy use and actual problem solutions; and bulletin board discussions. These data were complemented by further data gathered from small face-to-face focus group discussions.

Data have been gathered from focus groups giving feedback on technical issues and development of problem-solving skills. Two main conclusions have been gained from this evaluation. First, students have greatly appreciated the availability of other students' answers and particularly the comments on them. This has allowed students to put their own answers into a better perspective and to gain a clearer idea of what the lecturer expects. (The fact that assessment and lecturers' expectations drive the majority of students cannot be ignored.) Access to alternative solutions provides students with process models and supports reflection by the students on many aspects of problem solving. Coincidentally, other student answers help some students to feel more of a part of a group and that they are "not so stupid after all," which can affect student motivation to succeed along with others. Students have noted that their motivation often parallels their success in the subject. Becoming part of a community of students working in a course is particularly important for distance-education students, who may otherwise be rather isolated. The ability to discuss issues on the bulletin board has been helpful for these students.

Second, students have mentioned that planning and analysis of problems, whereby parts can be tackled step by step, has been helpful to them. Becoming more aware of and practicing such skills has been beneficial for students who in the past may have been intimidated by a problem and may have given up too soon when it seemed too difficult. Although some models and coaching online have been provided, students appear to need a more staged approach, with step-by-step examples and support. This "cognitive apprenticeship" approach is in tune with the constructivist principles underpinning the design of metAHEAD. Box 4.1 shows a summary of student comments gained from the evaluation.

Technical difficulties present the greatest obstacles to the ease of use of the tutorial at this time. To be able to solve problems, students need to be able to draw diagrams (in all subjects) and use mathematical formulas and equations (particularly for chemistry and biophysics). To carry out these tasks in any computer environment requires a considerable learning curve for most students at present. Moreover, while students can easily solve problems on paper employing diagrams and/or formulas, the online logbook, where they keep a record of their progress over the semester, can handle plain text only. The ability to keep an adequate record of a student's progress over a semester in an online format is a major design feature if a more student-friendly, computerized environment is to be provided.

CONCLUSION

Students' metacognitive skills can be developed significantly by taking a proactive approach and by designing learning environments specifically for problem solving and the development of metacognition. The choice of problems students are exposed to is vitally important for the development of higher-level cognitive skills. Learning environments need to be developed in contexts that engage students in self-monitoring their problem-solving approaches in scenarios where they can ultimately use that knowledge. This requires creating real-life anchors for development of problem-solving skills and enabling students to explore, test, and review their own strategies.

These learning environments do not necessarily need to be computer-based. At the moment the push for ICT-supported learning

---

**BOX 4.1. Student Comments Gained
from Evaluation of metAHEAD**

Comments on what students gained from metAHEAD

- Helps think about level of confidence and predict degree of success.
- Helps motivation somewhat.
- Helps with ways to understand the question better in order to tackle it successfully.
- Helps with breaking up tasks—something they didn't commonly consider explicitly.
- Talking to other students is helpful, hear other views.
- Getting feedback from lecturers, hearing them explain and comment on lecturers' answers.
- Appreciated other students' answers, so they could compare their own answers. Lecturers' "metacomments" on these were important.
- Other students' answers also show other ways of solving problems.
- Information about learning and thinking processes—e.g., chunking information.

Comments on limitations of metAHEAD

- Can't replace face to face. Need personal instructor for motivation.
- Needs to help students with more transfer to new problems.
- Needs more models and demonstrations of how to work through problems step by step.
- Logbook tool has great technical deficiencies, particularly for science answers involving formulas and equations.
- Audio and video quality needs to be better.

---

comes about partly through a need to accommodate distance-education students, as well as the need to teach larger student numbers with decreasing staff numbers. In fact, using ICT to assist the development of higher-order cognitive skills represents a very significant challenge. A face-to-face encounter of a learner with an experienced personal human instructor who has expert subject knowledge and pedagogical knowledge bases and is able to react flexibly to different situations is still far superior to the best computer tutors. It will require that much more of the tacit knowledge of expert teachers be made explicit and then incorporated into the knowledge databases of intelligent computer tutors before ICT can seriously replace face-to-face teaching. Significant technical difficulties also limit the convenience of using computer environments for writing out and depicting solutions to the sorts of problems students tackle in tertiary science.

The development of technology-supported resources is costly and time consuming, and in-depth evaluation is essential not only for design but also for development and implementation of such programs. An initial evaluation has shown both positive and negative aspects of the metAHEAD tutorial, and insights gained will be used to improve the design of the resource and its capacity to support problem solving. The choice of evaluation approach has been effective and worthwhile, providing valuable data on context, interactions, and outcomes of the tutorial. Continuing evaluation will assist in further refinement, tailoring the tutorial to student needs and improving its relatedness to the support of metacognition in tertiary-level science studies.

REFERENCES

Alexander, P.A. and Judy, J.J. (1988). The interaction of domain-specific and strategic knowledge in academic performance. *Review of Educational Research,* 58(4): 375-404.

Boekaerts, M., Pintrich, P.R., and Zeidner, M. (2000). *Handbook of self-regulation.* San Diego, CA: Academic Press.

Clarke, R.E. (1992). Facilitating domain general problem solving: Computer cognitive processes and instruction. In E. De Corte, M.C. Linn, and H. Mandl (Eds.), *Computer-based learning environments and problem solving* (pp. 265-283). Berlin: Springer Verlag.

Elen, J. and Lowyck, J. (1999). Metacognitive instructional knowledge: Cognitive mediation and instructional design. *Journal of Structural Learning and Intelligent Systems,* 13(3): 145-169.

Everson, H.T. and Tobias, S. (1998). The ability to estimate knowledge and performance in college: A metacognitive analysis. *Instructional Science,* 26(1-2): 65-79.

Gabel, D. (Ed.) (1994). *Handbook of research on science teaching and learning.* New York: Macmillan Publishing Company.

Gourgey, A.F. (1998). Metacognition in basic skills instruction. *Instructional Science,* 26(1-2): 81-96.

Gredler, M.E. (1997). *Learning and instruction: Theory into practice.* Merrill, NJ: Prentice-Hall.

Hatano, G. and Inagaki, K. (1986). Two courses of expertise. In H.A.H. Stevenson and K. Hakuta (Eds.), *Child development and education in Japan* (pp. 262-272). New York: Freeman.

Hobden, P. (1998). The role of routine problem tasks in science teaching. In B.J. Fraser and K.G. Tobin (Eds.), *International handbook of science education*, Volume 1, Part 1 (pp. 219-231). Dordrecht, the Netherlands: Kluwer Academic Publishers.

Hollingworth, R.W. and McLoughlin, C. (2001). Developing science students' metacognitive problem-solving skills online. *Australian Journal of Educational Technology,* 17: 50-63.

Johnstone, A.H. (1998). Learning through problem solving. In D. Rafferty and S. Sleigh (Eds.), *Problem solving in analytical chemistry* (pp. v-viii). London: The Royal Society of Chemistry.

Jonassen, D. (1997). Instructional design models for well-structured and ill-structured problem-solving learning outcomes. *Educational Technology Research and Development,* 45(1): 65-94.

Jonassen, D. (2000). Toward a design theory of problem solving. *Educational Technology Research and Development,* 48(4): 63-85.

Jonassen, D. (2002). Learning to solve problems online. In G. Glass and C. Vrasidas (Eds.), *Current perspectives on applied information technologies.* Volume 1: *Distance learning* (pp. 75-98). Macomb: Center for the Application of Information Technologies, University of Western Illinois.

Jones, A.E.S., Tosunoglu, C., and Ross, S. (1996). Evaluating CAL at the Open University: 15 years on. *Computers in Education,* 26(1): 5-15.

Lajoie, S.P. (1993). Computer environments as cognitive tools for enhancing learning. In S.P. Lajoie and S.J. Derry (Eds.), *Computers as cognitive tools* (pp. 261-288). Hillsdale, NJ: Lawrence Erlbaum.

Lin, X., Hmelo, C., Kinzer, C.K., and Secules, T.J. (1999). Designing technology to support reflection. *Educational Technology Research and Development,* 47(3): 43-62.

Lin, X. and Lehman, J.D. (1999). Supporting learning of variable control in a computer-based biology environment: Effects of prompting college students to reflect on their own thinking. *Journal of Research in Science Teaching,* 36(7): 837-858.

Lowe, I. (1999). Education for a rapidly changing future. Available online at <http://www.aisq.qld.edu.au/publications>.

Masui, C. and De Corte, E. (1999). Enhancing learning and problem-solving skills: Orienting and self-judging, two powerful and trainable learning tools. *Learning and Instruction,* 9(6): 517-542.

Mayer, R.E. (1997). Incorporating problem solving into secondary school curricula. In G.D. Phye (Ed.), *Handbook of academic learning: Construction of knowledge* (pp. 474-492). San Diego, CA: Academic.

Metcalfe, J. and Shimamura, A.P. (1994). *Metacognition: Knowing about knowing.* Cambridge, MA: MIT Press.

Oliver, M. (1999). ELT report 1. A framework for evaluating the use of educational technology. Available online at <http://www.unl.ac.uk>.

Schraw, G. (1998). Promoting general metacognitive awareness. *Instructional Science,* 26(1-2): 113-125.

Shin, N., Jonassen, D.H., and McGee, S. (2003). Predictors of well-structured and ill-structured problem solving in an astronomy simulation. *Journal of Research in Science Teaching,* 40(1): 6-33.

Taconis, R., Ferguson-Hessler, M.G.M., and Broekkamp, H. (2001). Teaching science problem solving: An overview of experimental work. *Journal of Research in Science Teaching,* 38(4): 442-468.

Volet, S.E. (1991). Modeling and coaching of relevant metacognitive strategies for enhancing university students' learning. *Learning and Instruction,* 1: 319-336.

Volet, S.E., McGill, T., and Pears, H. (1995). Implementing process-based instruction in regular university teaching: Conceptual, methodological and practical issues. *European Journal of Psychology of Education,* 10: 385-400.

Wolfskill, T. and Hanson, D. (2001). LUCID: A new model of computer-assisted learning. *Journal of Chemical Education,* 78(10): 1417-1424.

Zeegers, P., Martin, L., and Martin, C. (1998). Using learning to learn strategies to enhance student self-regulated learning in first-year chemistry. Paper presented at the Third Pacific Rim Conference, Auckland, New Zealand, July 5-8.

Zoller, U. (2000). Teaching tomorrow's college science courses—Are we getting it right? *Journal of College Science Teaching* (May): 409-414.

Chapter 5

Distributed Problem-Based Learning and Threaded Discourse

Lisa Lobry de Bruyn

INTRODUCTION

The use of problem-based learning (PBL) as a teaching and learning strategy has been well accepted in vocational degrees such as medical sciences, education, and law. However, the use of problem-based learning in the natural sciences has been more recent and has experienced varying degrees of success (Harland, 2002; Lobry de Bruyn and Prior, 2001a,b; Trevitt and Sachse-Akerlind, 1994). An even more recent development has been that of online instruction, and the use of computer-mediated communications in the delivery of distributed problem-based learning (dPBL) exercises.[1] Both educational developments, use of problem-based learning and online delivery of units, have arisen from the perceived need to diversify university teaching approaches to produce more competent graduates who can support the rapid changes occurring in the workplace. Since graduates are expected to have a range of skills and competencies along with the knowledge base to support them in the workplace, the natural sciences have often focused on the problem-solving aspect of problem-based learning to teach students about conflict resolution, negotiating change, and working cooperatively in teams (Touval and Dietz, 1994). In addition, PBL is an excellent teaching strategy that has the ability to strengthen and develop student competencies in the areas of information literacy, communication, self-directed learning, and solving "real-world" problems (Boud and Feletti, 1997). PBL is a learning approach that allows the integration of knowledge acquisition and teaching strategies to actively engage students in the learning process. Furthermore, online delivery offers distance students the ability to communicate asynchronously with other students to brainstorm, analyze, and redefine the learning issues for the problem-based learning exercise as well as social interaction. Nevertheless, there remains some reticence among students and educators regarding how well the learning strategy and delivery mechanism (problem-based learning and online delivery, respectively) combine. Hence the goals of this chapter are to

1. examine the changes required to the learning environment in order to deliver dPBL;
2. evaluate how effectively dPBL has been in developing a collaborative learning environment and student competencies in the areas of information literacy, communication, self-directed learning, and solving real-world problems;
3. examine strategies and ideas that could improve students' performance and experience of dPBL; and
4. examine the implications for the teacher of supporting dPBL, especially in courses with large numbers of students and/or with a large proportion of students studying off-campus.

BACKGROUND AND LITERATURE REVIEW

The literature review provides an overview of three interconnecting aspects: first, the traditional model of problem-based learning and its learning environment; second, how the model of PBL becomes modified with the advent of dPBL; and third, the benefits and drawbacks of using asynchronous, computer-mediated communication (ACMC) for delivery and exploration of problem-based learning exercises.

Numerous researchers have established the value of traditional PBL to learning (Boud and Feletti, 1997; Savin-Baden, 2000). In most situations the use of PBL is invaluable to instill in students the value of knowledge building and acquisition. Savin-Baden (2000) argues that problem-based learning in the arena of higher education has been misunderstood, interpreted too narrowly, and utilized in limited ways, and therefore educators have not realized its full potential as a learning approach. One of the issues mentioned by Savin-Baden (2000) is that in higher education the "focus on skills development at the expense of the development of abilities to research and critique information" (p. 15) has lead to a decline in students' ability to think and reflect on the information presented to them, and the significance of that information. The model of problem-based learning should combine "know how" with "know that" so that there is a greater synthesis of skills and knowledge acquisition and application and greater emphasis on combining the "world of work" with the academic context, without diminishing the value of either.

The traditional model of problem-based learning that the study reports on is expounded by Barrow (1985, 1988, 2002), and has four key aspects:

1. The problems are presented to the learner as real-world situations that are "unresolved ill-structured problems," which then stimulate the learners to generate questions about what has occurred and how they would respond. The situations are designed to motivate the learners to gather further information to resolve and understand the situation.
2. The learners take on responsibility for their learning and determine their learning needs in terms of information and locating appropriate and relevant resources to assist them in solving the problem, hence problem-based learning is a learner-centered pedagogy.
3. The teacher acts as a facilitator of learning, not a "sage on the stage," and encourages students to be autonomous learners and conduct self-directed research.
4. The problems selected are those most likely to be encountered by the learner in the "world of work," and the skills and activities required by the learners to solve the problems are also valued by the real world, making PBL an authentic learning process.

A special issue of *Distance Education* (2002, 23[1]) published a number of articles exploring as well as developing models for distributed problem-based learning. As more courses are being delivered by distance education and online mode due to recent educational developments in higher education (such as large class sizes and more students studying remotely) the efficacy of dPBL in an online environment has been questioned. The reservations about dPBL are often about the type of learning environments that are being created and the tools available to support communication when PBL is distributed in a virtual environment (Orrill, 2002). Even though the technology is now available to combine computer simulations of authentic problems for learning, it appears the technology is being underused or is not producing the predicted benefits to learners and educators (Lehtinen, 2002).

Central to the traditional model of problem-based learning is group work. The various steps of problem-based learning are conducted face to face in small groups: introducing one another, setting ground rules, acknowledging prior learning, identifying contributions to group learning, and working through the problem-solving process. When transferring traditional PBL to an online environment, ACMC is one avenue that allows students to communicate independently of time and place, and even accommodating small groups to be created to discuss questions, opinions, and queries. It seems those researchers who report positive outcomes using an electronic learning environment for problem-based learning are using it in addition to face-to-face sessions (Ronteltap and Eurelings, 2002), not instead of.

Focusing solely on the use of ACMC for supporting dPBL, a number of benefits are cited by researchers (Harasim et al., 1998; Hewitt, 2001; Mason and Kaye, 1989, 1990), including the following:

1. *Connectivity and accessibility:* There is increased group interaction since the discussions are open and not limited to face-to-face meeting times. Also, the collective knowledge of the class and other outside links are more accessible to students (Eastmond, 1994).
2. *Equitable communication* between students is encouraged, as there is no need for turn taking (Graddol, 1989), and everyone can be heard, including those more reticent students, without being intimidated by more vocal students.
3. *Student reflection* is also fostered by messages being preserved electronically, which can be revisited and reread later, allowing time for reflection before committing their own ideas to public scrutiny (Mason and Kaye, 1990).
4. *Boundless conversation:* Student conversations using ACMC are boundless in time and space and promote greater student interaction. Also, because time and location do not restrict com-

munication, the expectation is that all students will contribute to discussions.

With dPBL the notion of creating small groups of students as active, reflective participants in an electronically linked community is the ideal and not necessarily the reality. Researchers have summarized some of the types of issues that can be expected when using ACMC to allow discussion on learning issues (Guzdial and Turns, 2000; Harasim et al., 1998; Light and Light, 1999). For students some possible drawbacks may include the following:

1. Computer technical problems result in lack of student access to computer hardware or software.
2. Communication anxiety among students, especially in asynchronous environments when responses are not immediate. Also, because postings cannot be erased, students are concerned about the permanency of an ill-conceived message and how it will be perceived by other students/instructor. Also, those students who are new to the online environment may be reluctant to get the conversation flowing or may lack confidence in the online medium.
3. Limited student interaction, either because the learning environment does not motivate students to interact as it is unfamiliar or because of low student confidence in contributing new ideas to the discussion (Guzdial and Turns, 2000).
4. The lack of support for convergent (e.g., analyzing and synthesizing) processes (Hewitt, 2001).
5. Time management is often necessary, as the time spent online can easily exceed face-to-face classes. Online discussions are boundless (time and location) and are typically open-ended.
6. Information overload can occur due to the amount of information and links to other material, which can be placed on an online unit and can overwhelm students.
7. Misconceptions can occur when using threaded discussion groups, as there is sometimes no clear feedback to students to indicate if their point is clear. This situation is further compounded by "learner reluctance to push peer thinking and understanding" (Hewitt, 2003).

8. Traditional roles are often maintained if "the student speaks, the teacher answers, confirms, approves and reinforces" (Henri, 1995, p. 158, quoted in Light and Light, 1999). This form of teacher-directed dialogue discourages student interaction.

DESCRIPTION OF LEARNING ACTIVITY

The example of dPBL and use of ACMC in the present study is drawn from a course taught to a mixed-degree, dual-mode student cohort, concentrating on those students learning off-campus and utilizing hard copy and online teaching materials. The course is titled Land Evaluation and Land Degradation and is taught to students at the third-year university level. Characteristically, these students are learning at a distance (off-campus), are of mature age and already employed, and are undertaking the unit at bachelor's, graduate diploma, or master's level.

The reasons for introducing PBL into the curriculum was to immerse students in the curriculum and to assist them to understand the process of knowledge building as well as the disciplinary content of the knowledge (see Table 5.1). Many undergraduate students on completion of their degree will find work in land management agencies either at local, state, or federal level, and will often find themselves working in teams with disparate backgrounds and experiences in problem solving. Graduates must function effectively as group members as well as achieve progress within their working project. Thus, they must understand how to work as part of a team, delegate tasks, make joint decisions, and allocate resources. Hence, PBL involves students in learning about teamwork and implementing a range of tasks to achieve an outcome (e.g., interpersonal skills, time management, report writing, communication, and active listening) (see Table 5.1). In addition, the other desired learning outcomes of using problem-based learning were problem solving; information literacy (i.e., the ability to access, read, synthesize, and interpret information); alignment of content, class activities, and assessment tasks; fostering student motivation; and acknowledging prior learning and encouraging "intellectual prospecting" (Lobry de Bruyn and Prior, 2001a).

TABLE 5.1. Example of Problem-Based Learning Situation Statement

Student activities	Examples	Student skills and competency development areas
Students confronted with an ill-structured problem	The crop was resown, in parts, where the newly germinated plants had been washed away. John observed that in other areas, which were unaffected by the minor flooding, the plants did not grow vigorously. He pulled out a plant in the area where they were stunted in form and noticed there was poor root development. It is now near the end of the season, and the oats are being harvested. John had noticed earlier in the season that parts of the pasture—Lucerne—were yellow, as were parts of the oat crop. The yellowing seemed to be in patches. The oats will be baled for hay. The harvester had a yield monitor installed and John's records showed patchy yields, particularly in the lower parts of the landscape.	Information literacy, comprehension, and interpretation skills
Students requested to answer questions posed at the end of situation statement	What is happening to the soil? What soil health problems does John have? How should John respond?	Written expression, research skills, comprehension, and interpretation skills
Student discussion of problem in ACMC environment		Participation, leadership, team-building, and interpersonal social and communication skills
Students raise questions in response to situation statement questions	What are the soil types on the property? What soil tests have been completed on the property? Do a soil test for nutrients, soil pH, organic carbon. [How long fertilizer has been used on the farm] Did he fertilize when he resowed? What is the water table level? How long between flooding and resowing of seed before machinery used?	Higher-order thinking, problem solving, and interpersonal social and communication skills

TABLE 5.1 *(continued)*

Student activities	Examples	Student skills and competency development areas
Students seek information to solve problem	Examine soil testing data supplied. Investigate properties of soil types and eliminate the soil health problems that are not supported by evidence. Gather information on most optimal responses to soil health issues that are adoptable, practical, and relevant to situation.	Adoption of greater responsibility for learning goals and achievements, information literacy skills
Students derive multiple solutions and compose most plausible and likely answer for assessment		Creativity, written communication, and fluency skills

ASSIGNMENT STRUCTURE AND LEARNER SUPPORT

The use of distributed problem-based learning (dPBL) with off-campus students was first tested in 2000 with a small group of students who chose the alternative assessment ($n = 12$), while on-campus students had been undertaking PBL since 1999. In the following years (2001, 2002) the dPBL exercises were made a compulsory part of the course assessment.

The dPBL assignment is structured around the course curriculum, which focuses on identification of causes and solutions to land degradation problems and the concepts and practices involved in land use planning. DPBL in the Land Evaluation and Land Degradation course with off-campus students is applied in the following way. First, introductory notes on PBL are given to the students (available online and in hard copy). These include information on: What is PBL? How is it different to subject-based learning? and How would it be utilized in the course? PBL scenarios (retitled "situation statements") are introduced to the students in weeks 1, 4, and 10 of the course via the course home page or read as hard copy (example shown in Table 5.1). The dPBL assignment is completed in three stages: stage 1 and 2 (30 percent of course grade) are submitted together two-thirds (week 9) through the course, while stage 3 (30 percent of course grade) is submitted at the end of the semester (week 13). Assessment of student performance is based on a criterion-referencing framework using a marking schedule. On the bulletin board, the instructor emphasizes that students will be assessed on the realism of their answers and their ability to advocate a solution rather than provide a "right or wrong" answer. The key features of the assessment framework are that it is based on observable/demonstrable performance criteria, avoids jargon, and provides focused feedback on what students have done well, where they can improve, and aspects they should reconsider. Students are also asked to reflect by addressing the following question: "How confident are you that your answer is the best one? Explain why."

Students (on- and off-campus) are provided with scaffolding in the form of a structured learning guide to questioning and stimulating the problem-solving process (see Appendix). The three-step problem-solving process enables them to identify what they already know about the situation statement, what they need to know, and what they need to do in order to arrive at solutions to the questions posed at the end of the situation statement (see Table 5.1). For each situation an individual student response is to be written that addresses the situation statement questions in the few weeks that followed its introduction. Student participation is encouraged, especially for the problem-solving part of the dPBL exercises, by using a bulletin board (on WebCT, a leading course management system for facilitating online learning) and self-directed learning activity (i.e., reading and research), either using the online teaching material or material in addition to that supplied. Students are expected to follow the structured learning guide (see Appendix) and to discuss questions brought up by peers on the bulletin board at set times. Due to the number of students enrolled in the course (usually seventy) it was stated at the beginning that responses posted on the bulletin board would be to the student collective and not to individual postings. On the bulletin board, a week after "meet the situation," a rejoinder was posted by the instructor that attempted to "flesh out" with information the questions asked by students on the bulletin board. It was emphasized, in instructor postings, that students needed to either conduct further reading outside online teaching materials or utilize the material supplied. Working with on-campus students, the learning environment for PBL was typically characterized by students working in small groups, interacting and discussing issues face to face in a classroom. The study sought to examine the effects on the learning environment and student mastery when the same scenario statements were distributed online, with the initial student discussions of the PBL exercises conducted via asynchronous online discussion.

A number of strategies to support learner adjustment to dPBL were employed. One of the main strategies was to remove ambiguities in the learning task. This was achieved by recasting the language of the problem-based learning exercise so that students avoided literal translations, providing scaffolding of student learning with face-to-face sessions, and a "trial run" of a problem-based learning scenario using a case study outlined by Savoie and Hughes (1998). Schmidt and Moust (2000) report that the quality of the problem exerted the most influence over group functioning, as well as time spent on self-directed learning and interest in the subject matter. To reduce student anxiety about how to respond to situation statements and what to do with the information provided in response to their questions, the situation statements will use hot links on key words to link students directly to other Web sites, either to explain the term or to act as a catalyst for independent research activity. To redirect student focus onto their individual and collective responsibility, the decision was made to establish at the beginning a set of ground rules that explicitly identify student responsibilities to identify instructor, student, and group actions that would support learning (Lobry de Bruyn and Prior, 2001b).

EVALUATION OF LEARNING ACTIVITY
AND STUDENT/INSTRUCTOR EXPERIENCES

The strategies used to assess the success of the dPBL exercises' learning outcomes were summative evaluation, evaluation of assessment via student grades, and evaluation of student use of the bulletin board for problem solving and degree of convergent processes (i.e., degree of analysis, synthesis, and summarizing). Two classifications were used to analyze student postings in 2001 and 2002 for stage 1 of dPBL exercises, where most of the bulletin board discussion took place: one by Orrill (2002) and the other by Hewitt (2001), to respectively examine content of postings and the level of convergence occurring in postings (see respective tables for more detailed definition of classification schemes).

At the end of the semester students were asked for feedback on assessment and the use of PBL, and directly asked: "How do you think *you* could improve on *your* performance in the problem-based learning exercises?" Over half the students (57 percent) in 1999 responded by saying, for example, "by understanding what is required" or "what was expected to begin with, i.e., the marking schedule—had no idea what format or structure was expected" and "I need to have more specific knowledge of what the lecturer wants . . ." A few students identified their contribution to their performance by saying, "make more concrete suggestions/recommendations and display insights not regurgitated from text." The remaining students made no comment. In 2000, there was some improvement in students' appreciation of their responsibility to improve their performance in PBL exercises, with a 32 percent reduction in the proportion of students citing the level of explanation, guidance, or direction given by the instructor as a reason for their underachievement. In 2000, around 50 percent of the students realized by doing more research, reading, or expending more effort on their part they could have improved their performance in PBL exercises, while in 1999 only a few students had identified their contribution to PBL as affecting their performance. Also, a high proportion of students in the 2000 cohort (46 percent response rate) moderately to strongly agreed with both statements that they had improved their ability to solve problems (76 percent) and to think critically (57 percent). Upon subsequent researcher/instructor reflection it was decided that more scaffolding was required, especially in the first few weeks of the course, to facilitate students to take greater responsibility for their learning outcomes. In particular, students relied on the instructor to provide them with all the information required to solve the problem, rather than conducting their own self-directed research. Therefore, student behaviors and expectations that the instructor will perform the more traditional role of "sage on the stage" are difficult to alter.

Figure 5.1 shows the level of student mastery achieved in dPBL by examining the difference in grades for stages 1 and 2 (submitted two-thirds through the course) compared with stage 3 (submitted at the end of the course) over two years, 2001 and 2002. It appears, in both years, that the proportion of students achieving a distinction (D) and

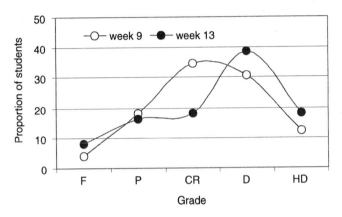

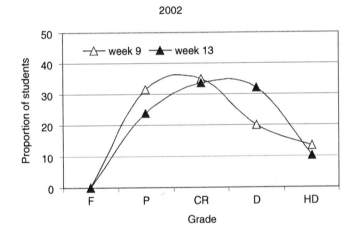

FIGURE 5.1. Student Results for Problem-Based Learning Assignment (as proportion of total students [n_{2001} = 49, n_{2002} = 60] over thirteen-week semesters in 2001 and 2002; grades correspond to following percentages: fail [F] < 50 percent; pass [P] 50 > 64 percent; credit [CR] 65 > 74 percent; distinction [D] 75 > 84 percent; high distinction [HD] > 85 percent).

high distinction (HD) improves (by 14 percent in 2001 and by 9 percent in 2002) over time. Also, when comparing student results between years, it is seen that no students failed any stage of the dPBL in 2002. Hence, the level of student competence, as measured by student achievement in dPBL, improved over the duration of the course and signified an increased student capacity to undertake dPBL.

The expectation that dPBL would create a sense of a learning community within an electronic environment by using discussion forums was not realized, especially with those students new to problem-based learning. Therefore, instructing off-campus students on how to approach PBL was more demanding, especially when participation was inconsistent, with unequal participation in posting comments. The use of a threaded, Web-based bulletin board that allows asynchronous communication is considered beneficial when student access is not predictable, and threading allows students to trace and keep track of conversational chains, as each note has a subject label and is organized in a hierarchical structure that includes only those messages that are related. Hence, unrelated threads are kept separate, and this allows students to pursue multiple avenues of thought without becoming confused (Hewitt, 2001). However, Hewitt (2003) also discovered that students are more likely to read the latest thread rather than read earlier threads.

Tables 5.2, 5.3, and 5.4 compare student bulletin board postings in 2001 and 2002 for stage 1 of the dPBL exercise. In 2001, the majority of students posted messages that combined task, problem solving, and other (38 percent), while in 2002 the messages focused more on problem solving and other tasks (35 percent) (see Table 5.3). In 2002, the proportion of problem solving only messages had increased, while the task only messages had declined in comparison to 2001

TABLE 5.2. Overall Assessment of Student Postings on First Stage of Problem-Based Learning Situation, 2001-2002

Factor	2001	2002	Change (2001-2002) in percent
Messages	37	48	30
Threads	22	24	9
Branches	12	26	117
Student postings	25	30	20
Instructor postings	3	4	33
Student repeat postings	9	14	56
Off-campus students in course	63	65	3

TABLE 5.3. Assessment of Content of Student Postings for Stage 1 of Problem-Based Learning Situation, 2001-2002 (student numbers; $n_{2001} = 25$, $n_{2002} = 30$)

Content of message*	2001	Postings/ student mean	%	2002	Postings/ student mean	%
Problem solving only	1	0.04	3	9	0.30	19
Task only	6	0.24	16	5	0.17	10
Other only	4	0.16	11	1	0.03	2
Problem solving and task integrated	0	0.00	0	0	0.00	0
Problem solving and task aspects	2	0.08	5	6	0.20	13
Problem solving and other aspects	8	0.32	22	17	0.57	35
Task and other	2	0.08	5	4	0.13	8
Task, problem solving, and other	14	0.56	38	6	0.20	13
Total	37		100	48		100

Note: Compiled using Orrill's (2002) classification scheme.
*Definition of classification scheme terms: problem solving = notes focused on presenting the students' thinking, asking a question related to the content of the problem, or otherwise engaged the students in thinking about the problem; task = notes were related to the logistics of the assignment task, such as verify layout, due dates, length of written component of assignment; other = notes did not focus on problem or the course.

postings. One hypothesis is that students in 2002 were better prepared to undertake dPBL than the 2001 cohort of students. Despite improvement in the volume of student postings (up by 30 percent), and student repeat postings (up by 55 percent) when comparing 2002 to 2001, the proportion of students that were involved in ACMC remained around 40 percent of the off-campus class for both years (see Table 5.2). Participation on the bulletin board was not part of student assessment, and although highly encouraged, was not compulsory to obtain a grade in the course. It seems a number of students were visiting the Web site and reading the messages on the bulletin board, but not contributing. Often this has occurred because they had missed the opportunity when the bulletin board discussions were occurring, for example, "I seem to be a little late with my questions and

TABLE 5.4. Assessment of Degree of Convergence of Student Postings on Stage 1 of Problem-Based Learning Situation for 2001-2002 ($n_{2001} = 25$, $n_{2002} = 30$)

Thread type*	2001	Postings/ student mean	% of total messages ($n = 37$)	2002	Postings/ student mean	% of total messages ($n = 48$)
Standalone	11	0.4	30	10	0.3	21
Add-on	26	1.0	70	39	1.3	81
Multiple	10	0.4	27	25	0.8	52
Convergence	0	0.0	0	0	0.0	0
Total	47			74		

Note: Compiled using Hewitt's (2001) classification scheme.
*Definition of classification scheme terms: Standalone = introduces new ideas to the conference and does not build on previous lines of inquiry. Typically, a standalone note is one that begins a new thread; add-on = builds on the ideas of one other note in the conference. Typically, these are notes in which one person responds to an idea that someone else has introduced; multiple reference = these notes make a reference to two or more previous notes, but not in a way that would be considered an attempt at convergence; convergent = a note that discusses some of the ideas expressed in two or more other notes in the conference.

most have already been asked!" Alternatively, students were reticent to post messages if they thought they were repeating what other students had already said, for example, "Hopefully not too repetitive," "I hope this is not repeating too much of what has already been said." Nevertheless, the positive outcomes of ACMC were clear from student postings, such as those that follow:

> "Really impressed with the questions raised. It really is a good means for cross-pollination of ideas and knowledge."
> "It's great to read everyone's questions."
> "A lot of good points have been made. I'll try and add a couple."

Two other ways of examining students' engagement in monitoring their own understanding are to examine their dispositions toward summarizing and their use of rationale for choices or decisions made. A posting is considered to use rationale if there is any opinion or evidence offered (Orrill, 2002). The results indicate that more than 70 percent of students in both years were posting merely add-on notes in response to postings of previous students, and there was no summarizing (see Table 5.4). However, when comparing 2002 with 2001, there was some improvement in student use of the bulletin board, with a higher proportion of multiple references to other students' postings, but still without convergence, and a decline in the proportion of standalone notes (Table 5.4). In Hewitt's analysis of student use of threaded online discussions, virtually all messages could be characterized as add-on notes, with few people attempting to tie together ideas from different sources. Hewitt (2001) suggested the reply convention of ACMC software prompts students to respond to a single note without considering the overall discussion (thread). Often, it seems students reply to a thread and leave the subject label (thread) unchanged, while the content of the message may have drifted away from the original purpose. It is also likely that students have not read earlier messages to examine how the discussion has evolved. Examining the use of rationale allows one to examine the level of involvement as well as whether the students are working collaboratively by explaining their position to others (Hewitt, 2003). Student postings in 2001 and 2002 in stage 1 showed that the use of rationale was not evident, and only after further prompting by the instructor did this improve.

CONCLUSION

There were a number of strategies used to improve students' experience and performance in dPBL. For students to value the learning skills that they were expected to use and develop through undertaking the dPBL exercises, these skills needed to be explicitly included in the assessment framework. Hence, independent research and learning skills were included in the marking framework (assessment rubrics), and students needed to demonstrate that they had conducted additional reading, research, analysis, and synthesis (all attributes of information literacy).

A number of suggestions to improve the use of ACMC as a tool to facilitate student inquiry, problem solving, and potentially improve level of convergence in bulletin board discussions were made by Hewitt (2001). These included appointing a moderator to summarize the discussion, preferably a student so other students could develop a deeper understanding of the problem-solving processes and the ways in which ideas may interrelate; assigning tasks that require group synthesis; using a linear discussion format;[2] and, finally, to augment ACMC with synchronous technologies (such as video conferencing), to make group coordination and negotiating group consensus easier. McLean (1999) also suggested separating the substantive content from "meta-communication" of the knowledge-building process to avoid cluttering the workspace with messages about due dates and deliverables rather than concentrating on the problems and issues under discussion. From the perspective of the study reported here, the use of ACMC could be improved by integrating student participation with assessment and learning outcomes, but the mechanisms of how to assess student interaction on the bulletin board needs to be carefully crafted to avoid an unwieldy and cumbersome assessment process.

Overall, the use of dPBL has identified some generic learning issues worthy of further research and application. These include the following:

1. Reducing student anxiety about the "fuzzy" nature of the PBL environment, where characteristically students are presented with "messy," ill-defined problems with better support and scaffolding of student learning "know how" and "know that"
2. Integrating the teaching of a range of problem-solving skills or tools (e.g., critical thinking, strategic planning, communication) along with or prior to the inception of dPBL exercises; for example, having the residential period (introducing and practicing PBL) prior to the start of the semester rather than two-thirds through the semester
3. Overcoming student preferences for face-to-face instruction on PBL rather than online instruction with asynchronous computer mediated communication, computer simulation, and hot links or print instruction
4. Examining further the interaction between the quality of the problem and student mastery
5. Creating sustained discussion on the bulletin board and cohesive learning groups when participation is voluntary and not heavily moderated
6. Incorporating frequent monitoring of students' learning progress and perceptions, increasing levels of communication with students concerning process and outcomes as well as ensuring a high degree of teacher responsiveness to student needs

APPENDIX

The learning guide is a significant scaffold/support provided to the students and is similar to the eight-step structured process expounded by Björck (2002), which he originally adapted from Barrows and Tamblyn (1980).
The learning guide includes the following:

1. Meet the situation (or scenario).
2. Define the situation.
3. Gather the facts.

   - Identify relevant experience and knowledge.
   - Identify what you need to know (further information and learning).
   - Identify potential information/learning resources (place ideas in 5).

4. Generate relevant questions from the previous section.

   - For you to go away and answer before next class
   - For me to go away and answer in next class

(Steps 1 to 4 are covered in class for the on-campus students. For the off-campus students, steps 3 and 4 are covered online via WebCT bulletin board discussion).

5. Research is required (type of…).
6. Rephrase the situation (refine the original question statement).
7. Generate answers (select possible, probable, and preferable explanations).
8. Advocate answers (choose the best answer and justify it).

Steps 6, 7, and 8 need to be written up and presented in an individual student six-page write-up.

NOTES

1. Distributed problem-based learning (dPBL), which has been defined by Cameron, Barrows, and Crooks (1999) as the use of PBL in online courses.
2. Linear discussion format does not allow branching and all notes are simply stored in a single, chronological order. WebCT bulletin board allows for threaded or unthreaded display of notes, with the latter being a linear discussion format.

REFERENCES

Barrows, H.S. (1985). *How to design a problem-based curriculum for the preclinical years.* New York: Springer.

Barrows, H.S. (1988). *The tutorial process.* Springfield: Southern Illinois Press.

Barrows, H.S. (2002). Is it truly possible to have such a thing as dPBL? *Distance Education,* 23(1): 119-122.

Barrows, H.S. and Tamblyn, R.M. (1980). *Problem-based learning: An approach to medical education.* New York: Springer.

Björck, U. (2002). Distributed problem-based learning in social economy—Key issues in students' mastery of a structured method of education. *Distance Education,* 23(1): 85-103.

Boud, D. and Feletti, G. (1997). *The challenge of problem-based learning.* London: Kogan Page.

Cameron, T., Barrows, H.S., and Crooks, S.M. (1999). Distributed problem-based learning at Southern Illinois University School of Medicine. In C. Hoadley and J. Roschelle (Eds.), *Proceedings of the Computer Support for Collaborative Learning (CSCL) 1999 Conference* (pp. 86-93). Mahwah, NJ: Lawrence Erlbaum Associates.

Eastmond, D.V. (1994). Adult distance study through computer conferencing. *Distance Education,* 15(1): 128-152.

Graddol, D. (1989). Some CMC discourse properties and their educational significance. In R. Mason and A. Kaye (Eds.), *Mindweave: Communication, computers, and distance education* (pp. 236-241). New York: Pergamon Press.

Guzdial, M. and Turns, J. (2000). Effective discussion through a computer-mediated anchored forum. *The Journal of the Learning Sciences,* 9: 437-469.

Harasim, L., Hiltz, S.R., Teles, L., and Turoff, M. (1998). *Learning networks: A field guide to teaching and learning online.* Cambridge, MA: MIT Press.

Harland, T. (2002). Zoology students' experiences of collaborative enquiry in problem-based learning. *Teaching in Higher Education,* 7: 3-15.

Hewitt, J. (2001). Beyond threaded discourse. *International Journal of Educational Telecommunications,* 7(3): 207-221.

Hewitt, J. (2003). How habitual online practices affect the development of asynchronous discussion threads. *Journal of Educational Computing Research,* 28(1): 31-45.

Lehtinen, E. (2002). Developing models for distributed problem-based learning: Theoretical and methodological reflection. *Distance Education,* 23(1): 109-117.

Light, P. and Light, V. (1999). Analyzing asynchronous learning interactions: Computer-mediated communication in a conventional undergraduate setting. In K. Littleton and P. Light (Eds.), *Learning with computers: Analyzing productive interaction* (pp. 162-178). New York: Routledge.

Lobry de Bruyn, L.A. and Prior, J.C. (2001a). Changing student learning focus in natural resource management education—Problems (and some solutions) with using problem-based learning. In L. Richardson and J. Lidstone (Eds.), *Flexible learning for a flexible society: Proceedings of ASET/HERDSA 2000 Joint International Conference* (pp. 441-451). Australia: ASET/HERDSA.

Lobry de Bruyn, L.A. and Prior, J.C. (2001b). Meeting of minds—Clashing of cultures: Evolution of teaching practice to engage students as co-learners. In P. Little, J. Conway, K. Cleary, S. Bourke, J. Archer, and A. Kingsland (Eds.), Annual HERDSA Conference 2001—Learning Partnerships (pp. 1-14). HERDSA: Canberra.

Mason, R. and Kaye, A. (Eds.) (1989). *Mindweave: Communication, computers, and distance education.* New York: Pergamon Press.

Mason, R. and Kaye, A. (1990). Toward a new paradigm for distance education. In L. Harasim (Ed.), *Online education: Perspectives on a new environment* (pp. 15-38). New York: Praeger.

McLean, R.S. (1999). Meta-communication widgets for knowledge building in distance education. In C. Hoadley and J. Roschelle (Eds.), *Proceedings of the Computer Support for Collaborative Learning (CSCL) 1999 Conference* (pp. 383-390). Mahwah, NJ: Lawrence Erlbaum Associates.

Orrill, C.H. (2002). Supporting online PBL: Design considerations for supporting distributed problem-solving. *Distance Education,* 23(1): 41-57.

Ronteltap, R. and Eurelings, A. (2002). Activity and interaction of students in an electronic learning environment for problem-based learning. *Distance Education,* 23(1): 11-22.

Savin-Baden, M. (2000). *Problem-based learning in higher education: Untold stories.* Buckingham: Open University Press.

Savoie, J.M. and Hughes, A.S. (1998). Problem-based learning as classroom solution. In R. Fogarty (Ed.), *Problem-based learning: A collection of articles* (pp. 95-102). Arlington Heights, IL: Skylight.

Schmidt, H.G. and Moust, J.H.C. (2000). Factors affecting small-group tutorial learning: A review of research. In D.H. Evensen and C.E. Hmelo (Eds.), *Problem-based learning: A research perspective on learning interactions* (pp. 19-51). Mahwah, NJ: Lawrence Erlbaum Associates.

Touval, J.L. and Dietz, J.M. (1994). The problem of teaching conservation problem-solving. *Conservation Biology,* 8: 902-904.

Trevitt, A.C.F. and Sachse-Akerlind, G. (1994). A district fire committee simulation in the professional forestry curriculum: A case study of computer-facilitated problem-based learning. In S. Chen, R. Cowdroy, A. Kingsland, and M. Otswald (Eds.), *Reflections on problem-based learning* (pp. 347-369). Newcastle, Australia: PROBLARC.

Chapter 6

Problem Solving in the Sciences: Sharing Expertise with Students

Catherine McLoughlin
Rowan W. Hollingworth

INTRODUCTION

In the context of higher education, the development of students' problem-solving skills in science continues to be an area of much ongoing research. Effective teaching of problem solving requires the adoption of process-based approaches that reveal to students the ways in which experts solve problems, as well as the coaching of students in higher-order skills that lead them away from a preoccupation with just finding a solution and toward building up a repertoire of problem-solving strategies. It is suggested that science educators need to model problem-solving explicitly by thinking aloud and demonstrating the skills we seek to develop in our students. To do this, teachers need to become more aware of the strategies and thought processes that are applied in problem solving in chemistry. Only then can teachers become colearners and share expertise with students, by demonstrating and modeling problem-solving strategies.

BACKGROUND

In tertiary education, there is an urgent need for professionals who can anticipate and predict problems and find solutions to the complex, interdisciplinary problems occurring in the real world. To meet this need effectively, educators must now reexamine traditional methods of teaching problem solving.

In discussing pedagogical approaches at the secondary school level, Hobden (1998) states that

> [Problem-solving] is a routine activity occupying a large proportion of curriculum time and plays a central role in student's experience of classroom life. From the first days of science instruction, sets of routine problem tasks assigned by the teacher have been part of classroom life. As a teaching strategy, they have largely been used uncritically. It would appear that nearly all physical science education, and especially the physics component, seems to be based on the optimistic assumption that success with numerical problems breeds an implicit conceptual understanding of science. (p. 219)

Although this was written about science education in general, it applies to the way problem solving has been taught in the past, particularly at first-year university-level science. Reif (1983) suggests that science teachers often behave analytically and systematically in their own subject area, but often do not do so in other domains and may tend to behave unscientifically with regard to teaching problem solving. The traditional method of showing worked-out examples followed by practice represents a rather primitive approach. This contrasts to the teaching of acting, music, or a sport, where the skills required for competent performance are analyzed into their component subskills, taught systematically, practiced, and integrated so that expert performance results. In addition, such skills are taught using apprenticeship models of teaching.

Constructivist models of learning can readily provide the basis for and means of supporting the metacognitive aspects of problem solving. Vygotsky's zone of proximal development (ZPD) is the theoretical grounding for the notion of scaffolding (Vygotsky, 1978). Scaffolding is a form of coaching or tutoring that helps learners accomplish tasks they cannot accomplish without assistance, therefore aiding in the construction of expertise in the tasks, engendering autonomous performance and confidence. The literature on scaffolding provides a variety of possible implementations that are relevant to the development of skills in problem solving, and enable teachers to share their expertise.

Scaffolding, especially as it is considered social in character, has affinity with the traditional apprenticeship model of learning and is essentially the support the master or teacher gives apprentices in carrying out a task (Winnips and McLoughlin, 2001). Social constructivism presents the knowledge as social and behavioral, and the learning process is itself social and interactive. Collins, Brown, and Newman (1989) offer the idea of cognitive apprenticeship to take up the particularly sociocognitive character of knowledge acquisition. This goes beyond the notion of a social scaffold (peers or instructors aiding in task performance) and allows for technology itself to play a role in scaffolding learning behavior. Collins and colleagues present the idea of apprenticeship as a model of instruction that supports the learner's cognitive processes, which is particularly relevant to pedagogies for teaching problem solving and assisting students to move from the stage of being novices to becoming experts in the problem-solving processes.

NOVICES AND EXPERTS

According to the Dreyfus model of skill acquisition, a person moves through five stages on a path from novice to expert performance (Dreyfus and Dreyfus, 1986).

- At the *novice* stage, where most beginning students are, the approach is to recognize objective facts and features relevant to the skill. Because novices have no experience, they rely on what are essentially "context-free" rules to guide them and judge their performance by how well they follow these learned rules.
- With some practical experience and scaffolding, the learners move to the *advanced beginner* stage, where they can recognize situational elements as well as the context-free components. Experience in concrete situations allows them to see similarities to previous examples they have encountered.
- At the *competent* stage, the number of context-free and situational elements proliferate and may become overwhelming. The person can plan and organize a situation by distilling the most important facts. To reach this stage may take two to three years.
- The *proficient* stage is reached when the person starts to move beyond analysis and is able to appreciate the salient features while allowing the others to fall into the background. Previ-

ously, decisions have been made by following rules, but now the learners may start to trust their intuition.

- *Experts* just do what needs to be done based on mature and practiced understanding, which derives from a vast storehouse of previously encountered situations. They are no longer governed by rules, since they understand the context and so know when and how to disregard, bend, or adapt their own strategies and skills.

Figure 6.1 shows the continuum of skill development from the rule-governed behavior of the novice to the context-driven behavior of the expert.

Of course, expertise is domain dependent and an expert scientist may be a novice in another area, for instance, teaching science. Because so much of the behavior of the expert is intuitive, such a person may have problems making explicit the strategies and context-free skills that the novice needs to rely on.

Science Educators As Experts

Let us take the view that teachers of chemistry are experts in their own domain. Assuming this is the case, how might they begin to teach more effectively? It is estimated that an expert in a domain has up to 50,000 chunks of working knowledge in his or her long-term memory and that it takes up to ten years to embed these chunks of knowledge and build connections between them (Schoenfeld, 1992). However, experts not only use their extensive knowledge database but also employ active strategies in problem solving. Experts are able to create mental representations of problems based on abstract principles, whereas novices are more likely to build representations based

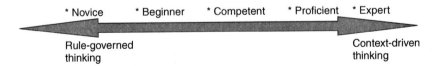

FIGURE 6.1. Progression Along the Stages of Skill Acquisition in Problem Solving

on more concrete and sometimes irrelevant features of the problem (Davidson and Sternberg, 1998). Successful problem solvers are also more likely to use a number of mental representations (Bodner and Domin, 2000) or to change their mental representations of a problem during its solution as a result of self-monitoring (Davidson and Sternberg, 1998). For example, Schoenfeld (1992) analyzed video recordings of the problem-solving activities of experts and novices in mathematics and produced revealing activity versus time graphs. When solving real problems students typically "read, make a decision quickly, and pursue that direction come hell or high water" (p. 62). This behavior usually guarantees failure, because the first quick decision is often wrong and is not subsequently changed. In contrast, an expert spends a large part of the time analyzing the problem and exploring possible solutions in a structured manner. Continual monitoring while trying interesting leads and abandoning false leads keeps the expert moving toward a successful solution. Science educators need to demonstrate the skills of problem representation, cognitive monitoring, and self-correction to their students in order to foster their expertise in problem solving.

As experts in the domain, science educators *may* make mistakes or make incorrect assumptions in dealing with problems, but they have well-developed monitoring and checking skills, which they can apply throughout the problem-solving process. Use of knowledge associations and constant checking provide teachers with many clues as to when the strategies adopted are appropriate and relevant and when a solution is heading up a blind alley. As a result of many years of experience, teachers (who are experts in their science domain) can apply skills and strategies automatically in solving problems but can less easily articulate these processes to their students.

It cannot be expected that students will be able to learn essential skills through short-term instruction or through teaching approaches that focus only on content and finding *the* solution. Effective teaching of problem solving requires modeling effective thinking processes and providing students with appropriate tasks. Problem-solving skills must be continually brought to the forefront and practiced, enabling students to become capable of more effective performance. For many science teachers this approach to teaching means radical change.

Student Misconceptions About Experts

Students may have misconceptions about the ways experts solve problems, and these misconceptions may arise through poor teaching. How often do teachers engage students and model problem solving by thinking aloud while solving an unrehearsed problem, demonstrating the actual thinking processes being used, making approximations, and attempting to correct errors explicitly in front of students (Woods, 1993)? Teachers are more likely to gloss over any mistakes or hesitations as quickly as possible, so as not to appear lacking in their area of expertise. Thinking aloud and demonstrating problem-solving processes may not be easy to undertake, because it requires an explicit awareness of and ability to verbalize the mental processes used to solve a problem. However, it is precisely this kind of modeling and verbal explanation of problem solving that students need to be exposed to. This process-based approach to teaching, if employed consistently with students, gives them the benefit of witnessing and observing expertise in problem solving. Teachers in the sciences need to demonstrate to students their expertise in using a variety of problem-solving strategies to help them become aware of expert problem-solving approaches and skills. The focus of curricula needs to be on the thinking skills and analytical strategies fundamental to problem solving.

In the past, much science teaching was done without reference to the findings of educational research, but educational practitioners now need to turn toward theoretically grounded and empirically supported teaching strategies with the aim of making teaching and learning more effective (Taconis, Ferguson-Hessler, and Broekkamp, 2001). Teachers need to recognize that students must move beyond the learning and application of facts, formulas, and inert knowledge to the development of higher-order thinking skills (Garratt, 1998).

DEVELOPING SELF-KNOWLEDGE

So how can experts in the field of science share their expertise with novices in the field? Since a significant part of teaching duties concerns supporting students, an effort must be made to reflect on teaching practice so that implicit knowledge becomes more explicit and accessible to the learner (Schon, 1995). One strategy is to focus on developing self-knowledge among learners through assessment and teaching strategies. Flavell (1976) proposes that self-knowledge is an important dimension of metacognition, and that includes knowledge of one's strengths, weaknesses, motivations, and learning strategies. Teachers can foster self-knowledge and reflection by creating learning environments where awareness and articulation of beliefs and motivational states are part of the learning conversation. Table 6.1 provides an overview of how self-knowledge can be supported by assessment and teaching strategies.

Teaching Practices Offering Support for Metacognition

The Role of Scaffolding

One form of scaffolding of relevance to development of problem-solving skills is the metacognitive scaffold. These usually take the form of self-monitoring, planning, and reflection prompts, which assist learners in mapping out their strategies for an activity and reflect back on that activity, identifying their work's strengths and weaknesses (Lin et al., 1999). These are different from mere process prompts, such as

TABLE 6.1. Examples of Self-Knowledge and How to Assess Them

Self-knowledge	Assessment and teaching strategies
Knowledge that one is knowledgeable in some areas but not in others	Helping learners become more aware and conscious of their own beliefs, and how to monitor those beliefs
Knowledge that one tends to rely on one type of "cognitive tool" (strategy) in certain situations	Compare his or her strategy with those used by other students
Knowledge of one's capabilities to perform a particular task that are accurate, not inflated (e.g., overconfident)	Self-assessment of success prior to undertaking the task
Knowledge of one's goals for performing a task	Articulation of goals prior to undertaking the task
Knowledge of one's personal interest in a task	Self-questioning and reflection on one's motivation for the task or problem
Knowledge of one's judgments about the relative utility value of a task	Self-questioning and comparison with peers' judgments on task utility

those that would attempt to teach an actual problem-solving process. Instead, these prompts attempt to invoke not so much cognition regarding the structure of a task or the particular knowledge domain, but rather metacognition, that is, "thinking about thinking." However, what needs support is some form of reflection by students, which would aid in the self-monitoring that is occurring while they engage in the processes of problem solving (McLoughlin, 1999).

Osman and Hannafin (1992) propose a number of strategies that can be used by teachers to support metacognitive strategies, as summarized in Table 6.2. Embedded strategies can be both content dependent and content independent. For example, the teacher may ask students to solve the problem, and therefore students learn to understand the content domain and the related problem-solving processes. However, unless the teacher provides an actual novel task, where students apply the strategies to other related content, then student skill development is limited.

Consider the following two problems concerning density, which might be given to students in entry-level university science.

1. A cube of metal weighs 1.45 kg and displaces 542 mL of water when immersed. Calculate the density of the metal.
2. Is the Aqua Ear product pictured in Figure 6.2 just a mixture of the two pure chemicals listed on the label or is it a water solution of those components? Use appropriate calculations to prove your answer.

TABLE 6.2. Embedded Detached Strategies for Metacognitive Development

	Embedded Strategies	**Detached Strategies**
Content-dependent strategies	Strategies taught with specific content; their use is constrained to a particular variety of content and the least amount of transfer is possible.	Strategies taught separately from content, but their utility is constrained to a particular variety of content; greater potential for transfer than embedded strategies.
Content-independent strategies	Strategies taught with specific content, but then become transferable to other content when a transfer task is provided.	Strategies taught separately from content; useful with a variety of content; greatest potential for transfer.

Source: Derived from Osman and Hannafin, 1992.

FIGURE 6.2. Aqua Ear Solution

Problem 1 is a typical decontextualized end-of-chapter exercise with all required data supplied. Students need only to insert data into a formula, D = m/V, to obtain the right answer without much thought. In contrast, Problem 2 needs transfer of skill to a novel situation. The problem is contextualized in relating to a consumer product students may use in their everyday life; it cannot be solved just by plugging numbers directly into an equation without thought. Students will need to search for extra data to solve the problem. Although this is in essence a very simple density problem, many students find it initially difficult to understand what the question is about.

   In adopting detached strategies to support problem solving, the teacher may again adopt content-dependent strategies or content independent strategies. The use of content-independent strategies produces the greatest likelihood of successful transfer, but, if the subject matter and procedural knowledge is not well-known to students, they might find this approach too challenging. The most productive approach is therefore to balance content-dependent and content-independent strategies, so that students progress from the security of tried and tested approaches, which are heavily contextualized, to more generic strategies that can be applied depending on task demands.

Questioning Strategies

In supporting students by engaging in think-aloud processes to foster metacognition and support problem solving, the teacher can make use of questioning strategies. By modeling the actual processes of problem solving while working through a particular task, students can observe the process of self-monitoring and then apply the same strategies while they engage in tasks. Essentially, this approach to teaching in the sciences requires a refocusing on the processes of thinking and a shift away from preoccupation with teaching facts, formulae, and inert knowledge. Table 6.3 provides examples of self-questioning strategies that can be used by teachers as they scaffold students' emerging skills in problem solving.

Changes in Teaching Problem Solving

Moving Away from a Focus on the Answer

As noted by Hobden (1998), students often in the past have been asked to solve routine exercises and teachers assumed that they

TABLE 6.3. Metacognitive Executive Processes Scaffold Prompts by Type

Type	Questions
Planning	• Do I have prior knowledge to help aid in this task? • What are the possible synonyms or alternative terms? • Where or how should I start? • How much time do I have?
Regulating (monitoring and revising)	• How am I doing? • Am I finding the information I need? • Am I in the right place? Should I try to search somewhere else? • What have I learned so far?
Evaluation	• Did I get what I was looking for? • Did I perform this task effectively? • What could I have done differently? • Are there some parts I still don't understand?

Source: Modified from Schraw and Dennison, 1994.

would develop problem-solving skills and strategies merely by exposure and practice. Problem-solving strategies have not usually been taught explicitly. In a typical teaching scenario, students would be shown how to solve problems in lectures and tutorials with the teacher demonstrating. The primary focus would be on the disciplinary content and then on finding *the* solution, rather than on helping students to develop independent problem-solving skills. Because the lecturer or instructor may have articulated the solution of problems in front of the class without thinking aloud, and because the "problems" were merely exercises rather than authentic, ill-defined problems, students were not exposed to the real processes involved in problem solving. In fact, students are often misled toward believing that the solution of problems always follows a *logical* process, that there is *one* correct way to devise the solution, and that lecturers do not make mistakes in solving problems. As a result, students then focus on model solutions rather than engaging with strategies that can be applied to a range of problems. Students taught in this way are likely to say "Just show us the answer," reflecting their preoccupation with content rather than process. There is a place for model answers or worked examples in learning problem solving, since they can reduce the cognitive load for beginners in a field. However, their use needs to be considered in the light of the previous discussion, and they cannot be relied on alone to develop problem-solving skills (Atkinson et al., 2000).

THE NEED FOR OPEN-ENDED TASKS

Are the problems given to our students, such as those appearing at the end of chapters of texts, really problems? Some would say these are merely exercises. Whether a particular task given to a student is a problem or not depends on the student, his or her knowledge, and his or her experience. Bodner and Domin (2000) suggest that "problem-solving is what you do when you don't know what to do, otherwise it's not a problem" (p. 22). Clearly a student at the beginning of tertiary study, faced with what might be regarded as a routine exercise, may genuinely view it as a problem, as he or she has limited domain knowledge. However, such tasks will no longer represent real problems to students after they practice a number of similar tasks.

Problems may range from simple to complex, from those where all the data required to solve them are supplied and which can be solved by a straightforward algorithm, to those which may lack data and are more ill-defined or open-ended (Reid and Yang, 2002). Solving more complex problems requires higher-order cognitive skills along the lines suggested in Bloom's taxonomy of educational objectives (Bloom et al., 1956), which has recently been revised by Anderson and Krathwohl (2001) (see Figure 6.3).

Too often the problems presented to students exercise only lower-order cognitive skills, and they are given too few opportunities to develop their higher-order skills by attempting more complex problems. Garratt (1998) lists some characteristics of authentic problems as engaging interest, connecting to prior knowledge, requiring decision making and judgment, and being open-ended or controversial. In order to develop more generally applicable problem-solving skills and strategies, pedagogical approaches need to ensure that students are exposed to a range of problems, from simple to complex and from well-defined to ill-defined.

Considering the Teaching of Problem Solving in Chemistry

Evidence of the changing emphasis in chemistry problem solving, for example, can be seen in the presentation of problem solving in textbooks. A decade or so ago, books devoted to chemistry problem solving essentially gave a brief summary of each topic, then sample work exercises, followed by many similar drilling exercises on each topic (Sorum and Boikess, 1981; Gibson and Faber, 1988). Little discussion of the processes of problem solving or the development of problem-solving skills was provided. Nowadays, most standard first-year general chemistry texts include some sections and tasks specifically related to problem solving (Kotz, Treichel, and Harman, 2003; Brady, Russell, and Holum, 2000). Often, the end-of-chapter exer-

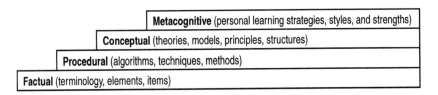

FIGURE 6.3. Anderson and Krathwohl's Knowledge Levels

cises in standard chemistry texts now include conceptual questions and challenging problems, going beyond the more routine problems and application of formulas.

In addition, more titles are appearing that deal explicitly with conceptual understanding, critical thinking skills, and problem solving (Barouch, 1997; Moog and Farrell, 1996; Bucat and Shand, 1996; Garratt, Overton, and Threlfall, 1999; Sleet et al., 1996). The emphasis in these publications is on the development of understanding through various types of thinking and problem-solving tasks, with less emphasis on quantitative calculation exercises. This process-based problem-solving approach assists students to actively develop problem-solving skills and discuss and explore their understanding, often in a collaborative manner.

RECOMMENDATIONS FOR TEACHING PROBLEM SOLVING

To restate the main point so far: Science teachers have tended not to effectively share their insights into expert problem-solving processes with their students. If teachers can develop awareness of their own problem-solving approaches, articulate these, and adopt student-centered approaches to teaching, they can start to genuinely share their expertise with their students. Zimmerman (1998) describes learning as "a proactive activity, requiring self-initiated motivational and behavioral processes, as well as metacognitive ones" (p. 4).

The teacher's role is to structure a developmental process that cultivates and respects alternative viewpoints, rewards innovative thinking, encourages awareness of cognitive growth, and requires defensible conclusions. The fear students feel about exposing their possible errors can be alleviated by a teacher who acknowledges the risk, gives learners permission to experiment, and encourages them to become confident in asserting their viewpoints, coupled with the responsibility for presenting a committed, substantiated position (Taylor, Martinau, and Fiddler, 2000).

Teachers need to recognize that the act of teaching has affective dimensions in the exploration of metacognition with the components of coconstruction of meaning, articulation of ideas, and role modeling. Each of these aspects is required for the student to be guided to the next stage of metacognitive awareness. Through a structured teaching approach that frees the vulnerable learner from a sense of personal deficiency, the learner is then able to begin moving toward reflecting on how his or her own actions have influenced the solution to the problem. Increasingly, students develop the ability to analyze their own reactions and alter their self-perceptions and strategies for problem solving.

Teachers can feel a sense of achievement when learners realize that it is possible to regulate their own thinking. The metacognitive analysis of the thinking behind choices, openly articulated by the teachers, allows examination of other problem-solving options. The additional layer of metacognition infused into teaching enables the learner to self-talk through a progression from "I am insecure because this situation is new to me and I am hoping to find an exact solution" to "I have choices in the strategies I can use and there may be more than one solution." The capacity to reflect on a range of possible approaches to solving the problem enables the student to consciously select from a widening repertoire of options and strategies and to feel a greater sense of agency and control.

Problem-solving strategies can be illustrated by initially applying them to routine exercises or structured problems as teachers think aloud and model the strategies employed. The preliminary analysis phase of the solution of an exercise happens very quickly for experts and it takes effort to verbalize what is happening. However, the initial qualitative understanding of the problem that the expert builds and the application of relevant principles and knowledge must be demonstrated. The process may be likened to producing a painting by first sketching an outline, elaborating on this sketch, and finally filling in all the details (Reif, 1983). The aim is to bring the student to a higher-level awareness of the processes they should use and to model more effective processes when appropriate. Habitual discussion with students about what they have learned from particular exercises can be beneficial if it does not focus solely on the solution but also on the thinking and planning processes of problem solving. Small group discussion with peer feedback also helps students to consider other perspectives. Helping learners to develop habits of reflection will lead to greater understanding of their own skills and more effective functioning as problem solvers. Table 6.4 provides a summary of process-based teaching strategies that can be adopted to foster effective problem solving.

TABLE 6.4. Process-Based Teaching Strategies to Foster Problem Solving in Science

Teacher action	Learner activity
Think aloud and articulate problem-solving strategies	Learner observes expert working through all problem-solving stages.
Model problem-solving strategies	Learner creates a mental model of expert problem solving.
Coach problem-solving strategies	Learner becomes an apprentice in problem solving and receives support from expert.
Give feedback by revealing learners' thinking processes to them as they solve problems	Learner can observe and assess his or her own problem-solving strategies through feedback.
Provide prompts and questions to stimulate learner reflection	Learner develops self-awareness and begins to self-monitor.
Plan opportunities for reflective peer discourse	Learner can engage in discussion with peers and experience multiple perspectives on problem solving.

CONCLUSION

This chapter has mainly been concerned with advocating the adoption of process-based pedagogies to develop problem-solving skills in science students. By systematically teaching students to reflect on and develop their problem-solving skills and by sharing metacognitive expertise with them explicitly, teachers can change students' approaches to learning. In this way, learners will be better equipped to tackle real, more open-ended, and more complex problems requiring higher-order cognitive skills. By adopting pedagogical activities such as thinking aloud, coaching, and articulation of problem-solving processes, teachers can assist students to develop a repertoire of skills and self-awareness of their own problem-solving strategies. Desirable skills can be fostered in science students by teachers, who have both awareness of expert problem-solving strategies and the capacity and willingness to share their expertise with students through their teaching.

REFERENCES

Anderson, L.W. and Krathwohl, D.R. (2001). *A taxonomy for teaching, learning, and assessing.* London and New York: Longman.

Atkinson, R.K., Derry, S.J., Renkl, A., and Wortham, D. (2000). Learning from examples: Instructional design principles from the worked examples research. *Review of Educational Research,* 70(2): 181-214.

Barouch, D.H. (1997). *Voyages in conceptual chemistry.* Sudbury, MA: Jones and Bartlett Publishers.

Bloom, B.S., Engelhart, M.D., Furst, F.J., Hill, W.H., and Krathwohl, D.R. (1956). *Taxonomy of educational objectives:* Handbook 1, *Cognitive domain.* New York: McKay.

Bodner, G.M. and Domin, D.S. (2000). Mental models: The role of representations in problem solving. *University Chemistry Education,* 4(1): 22-28.

Brady, J.E., Russell, J.W., and Holum, J.R. (2000). *Chemistry: Matter and its changes* (Third edition). New York: John Wiley and Sons.

Bucat, B. and Shand, T. (1996). *Thinking tasks in chemistry, teaching for understanding.* Perth, Australia: University of Western Australia.

Collins, A., Brown, J.S., and Newman, S.E. (1989). Cognitive apprenticeship: Teaching the crafts of reading, writing and mathematics. In L.B. Resnick (Ed.), *Knowing, learning and instruction: Essays in honour of Robert Glaser* (pp. 453-494). Hillsdale, NJ: Lawrence Erlbaum.

Davidson, J.E. and Sternberg, R.J. (1998). Smart problem solving: How metacognition helps. In D.J. Hacker, J. Dunlosky, and A.C. Graesser (Eds.), *Metacognition in educational theory and practice* (pp. 47-68). Mahwah, NJ: Lawrence Erlbaum Associates.

Dreyfus, H.L. and Dreyfus, S.E. (1986). *Mind over machine.* New York: The Free Press.

Flavell, J.H. (1976). Metacognitive aspects of problem solving. In L.B. Resnick (Ed.), *The nature of intelligence* (pp. 231-235). Hillsdale, NJ: Lawrence Erlbaum.

Garratt, J. (1998). Inducing people to think. *University Chemistry Education,* 2(1): 29-33.

Garratt, J., Overton, T., and Threlfall, T. (1999). *A question of chemistry: Creative problems for critical thinkers.* Harlow, England: Longman, Pearson Education Limited.

Gibson, G.W. and Faber, G.C. (1988). *Mastering chemistry problems* (Second edition). Englewood Cliffs, NJ: Prentice-Hall.

Hobden, P. (1998). The role of routine problem tasks in science teaching. In B.J. Fraser and K.G. Tobin (Eds.), *International handbook of science education,* Volume 1, Part 1 (pp. 219-231). Dordrecht, the Netherlands: Kluwer Academic Publishers.

Kotz, J.C., Treichel, P. Jr., and Harman, P.A. (2003). *Chemistry and chemical reactivity* (Fifth edition). Brooks/Cole, USA: Thomson.

Lin, X., Hmelo, C., Kinzer, C.K., and Secules, T.J. (1999). Designing technology to support reflection. *Educational Technology Research and Development,* 47(3): 43-62.

McLoughlin, C. (1999). Scaffolding: Applications to learning technology supported environments. Paper presented at the Ed Media 1999 World Conference on Educational Multimedia. Charlottesville, VA: Educational Multimedia and Hypermedia.

Moog, R.S. and Farrell, J.J. (1996). *Chemistry: A guided inquiry* (Preliminary edition). New York: John Wiley and Sons.

Osman, M.E. and Hannafin, M.J. (1992). Metacognition research and theory: Analysis and implications for instructional design. *Educational Technology Research and Development,* 40(2): 83-99.

Reid, N. and Yang, M.-J. (2002). The solving of problems in chemistry: The more open-ended problems. *Research in Science and Technological Education,* 20(1): 83-98.

Reif, F. (1983). How can chemists teach problem solving? *Journal of Chemical Education,* 60(11): 948-953.

Schoenfeld, A.H. (1992). Learning to think mathematically: Problem solving, metacognition, and sense making in mathematics. In D.A. Grouws (Ed.), *Handbook of research on mathematics teaching and learning* (pp. 334-370). New York: Macmillan.

Schon, D.A. (1995). *The reflective practitioner: How the professionals think in action.* Aldershot, United Kingdom: Ashgate Publishing.

Schraw, G. and Dennison, R.S. (1994). Assessing metacognitive awareness. *Contemporary Educational Psychology,* 19: 460-475.

Sleet, R., Hager, P., Logan, P., and Hooper, M. (1996). *Broader skill requirements of science graduates.* Sydney, Australia: University of Technology.

Sorum, C.H. and Boikess, R.S. (1981). *How to solve general chemistry problems* (Sixth edition). Englewood Cliffs, NJ: Prentice-Hall.

Taconis, R., Ferguson-Hessler, M.G.M., and Broekkamp, H. (2001). Teaching science problem solving: An overview of experimental work. *Journal of Research in Science Teaching,* 38(4): 442-468.

Taylor, K., Martinau, C., and Fiddler, M. (2000). *Developing adult learners: Strategies for teachers and trainers.* San Francisco: Jossey-Bass.

Vygotsky. (1978). *Mind in society: The development of higher psychological processes.* Cambridge, MA: Harvard University Press. (Original material published in 1930, 1933, and 1935.)

Winnips, K. and McLoughlin, C. (2001). Six WWW learner supports you can build. Paper presented at the Ed Media 2001 World Conference on Educational Multimedia. Charlottesville, VA: Hypermedia and Telecommunications.

Woods, D.R. (1993). Problem solving—What *doesn't* seem to work. *Journal of College Science Teaching,* 23(September/October): 57-58.

Zimmerman, B. (1998). Developing self-fulfilling cycles of academic regulation: An analysis of exemplary instructional methods. In D. Schunk and B. Zimmerman (Eds.), *Self regulated learning: From teaching to self-reflective practice* (pp. 1-12). New York: The Guilford Press.

Chapter 7

Student-Centered Learning Support in the Sciences

Robyn Muldoon

INTRODUCTION

The need for learning support in tertiary institutions is now an acknowledged and accepted fact in universities worldwide. This is largely because of increased participation in senior secondary and tertiary education, which has resulted in the enrollment at university of students from a wider range of backgrounds than previously. This has meant larger numbers of students who, for a variety of reasons, do not have the skills and background knowledge once considered essential for beginning university study (Baldauf, 1997; Garner, 1997; McInnes and James, 1995; Parker, 1997). No longer is it possible to assume that students are ready for specialized, academic study (Johnston, 2001; McInnes and James, 1995). Increasing numbers of first-year students are challenged by the expectations of tertiary institutions, with respect to thinking and learning, and many students struggle with their new role as independent learners and critical thinkers (Beasley, 1997; Johnston, 2001; McInnes and James, 1995). This is exacerbated by the recent proliferation of courses in response to increased demand and the introduction of new electronic forms of discourse (Taylor, 2000; Threadgold, Absalom, and Golebiowski, 1997). These days, the notion of tertiary literacy, which refers to the skills and abilities considered fundamental to success in and essential outcomes of tertiary study (Parker, 1997), is a significant issue.

Linked with tertiary literacy concerns is the wider issue of retention (McInnes, 2001). In Australia, 30 to 40 percent of students who begin university studies do not graduate, and the highest attrition occurs in the first year of study (Campus Review, 2003; Department of

Education Training and Youth Affairs, 2000). It is well known that initial experiences on campus are important and influence students' persistence in higher education (McInnes, James, and Hartley, 2000; Pitkethly and Prosser, 2001).

There is a great deal of evidence that students who are at risk of failing or dropping out are affected by a range of factors, not just tertiary literacy deficit (McInnes and James, 1995; Tinto, 1987; Yorke, 2001). Another of the previously held assumptions about first-year students that no longer applies is that students come from families and social environments which equip them to assume the lifestyle and meet the expectations of university life smoothly (McInnes and James, 1995). Increasing numbers of students need to balance their studies with part-time work in order to survive. Therefore, many students new to tertiary study need not only academic support but also emotional and moral support.

This chapter provides an overview of current concerns about the learning needs of tertiary science students and approaches to supporting learning and assisting students to meet the challenges of first-year study, particularly those challenges related to tertiary literacy deficit and those that result in student attrition. It also describes a specific student mentoring program, an effective amalgamation of those approaches which are currently being tested in the Faculty of the Sciences at the University of New England (UNE) in Australia.

SUPPORTIVE PEDAGOGIES
FOR STUDENTS IN THE SCIENCES

Expansion of the tertiary sector has had particular ramifications in science education in Australia and the United Kingdom. Students are now studying science for a range of different reasons and many arrive at university without having studied subjects once considered to be prerequisites to tertiary science. The explosion of knowledge in the sciences has resulted in tertiary science education tending toward qualifying students in the manipulation of this knowledge and basic scientific literacy rather than training the next generation of researchers, which is now largely a function of postgraduate study (Williams, 1991 in Laws, 1996; Solomon and Thomas, 1999). At the same time, student interest in science in general is declining (Niland, 1998). Not unsurprisingly, falling success and retention rates have been of particular concern in tertiary science education.

Many tertiary science teachers are searching for appropriate responses. Prosser and Trigwell (1999) believe that science students' previous experiences studying similar topics have a significant effect on the way they approach learning at the tertiary level. They recommend that an understanding of students' preconceptions and current needs are crucial to successful learning outcomes and that changing and adapting teaching and assessment (see Chapter 10) in relation to students' situations is the way to achieve a true student-focused approach. Others are questioning traditional teaching methods, such as lectures, as it is increasingly recognized that while they may reach large numbers of students, they do not necessarily guarantee student learning. Other traditional teaching methods, such as tutorials, are better suited to catering to a diversity of students' abilities and approaches to learning but by their very nature are not suited to large numbers. As Laws (1996) observes, "traditional teaching methods have become both necessary and anachronistic" (p. 3) and do not encourage the university science teacher wishing to improve the quality of student learning to adapt to student need, as recommended by Prosser and Trigwell (1999).

There has been a great deal of activity in Australian, British, and North American universities aimed at developing programs and support networks to increase success and retention rates of tertiary students, particularly those in the most problematic first or transitional year between secondary school and university. The result has been a dramatic growth in conferences and publications about specific interventions in the first-year experience. Ongoing themes of first-year experience conferences in Australia have been the generic versus discipline-specific learning support debate and the advantages and disadvantages of group work versus one-to-one learning support. Student mentoring, in its various forms, is an increasingly common approach to supporting the transition to tertiary study. This approach combines the known benefits of discipline-specific learning support with the best features of group work and one-to-one strategies. It is an ideal vehicle for not only assisting students to reach their full potential in tertiary study but also to achieve integration into the learning community and engagement with the institution, acknowledged ingredients in student retention (Tinto, 1998). Table 7.1 provides an overview of typical approaches adopted by universities in Australia for the provision of academic support to students.

TABLE 7.1. Summary of Approaches to Student Learning Support

Approach	Pedogogies used
Discipline specific	Teaching academic skills within the curriculum and related to real tasks and activities within courses of study: • Academic skills specialist giving lectures • Academic skills specialist advising teaching team • Collaborative teaching involving academic skills specialist and teaching team
Generic	Academic skills taught separately to the curriculum via broad-based topics applicable in any discipline, e.g., • Reading and note-taking skills • Effective study techniques • Basic essay writing
One to one	Teacher or mentor working with students on an individual basis. May be discipline specific or generic. Allows for personalized assistance with • Academic skills development • Individual barriers to learning • Transitional issues related to learning styles, the course of study, the institution, personal problems
Group work	Learning support occurring via collaborative learning situations. May be facilitated by teachers, academic skills specialists, mentors, tutors in residential college settings. Ideal for • Academic skills development • Lecture review • Exam preparation • Problem solving
Mentoring	Carried out by senior mentors or peer mentors via one-to-one or group work. May be discipline specific. Support may be emotional/psychological, academic, or a combination. Effective in assisting socialization into university life and increasing students' connection to the institution. Typical activities include the following: • One-to-one support as outlined previously • Generic and/or discipline-specific group work as outlined previously • Referring students to specialized assistance and support • Providing a friendly and nonjudgmental ear

Discipline-Specific Learning Support

For many years academic support programs in tertiary settings in Australia have been generic in nature and seen as remedial in intent. Academic support units have provided stand-alone workshops and short courses in a range of traditional study skills such as reading, listening, and note-taking skills. In the 1980s this began to change, as academic support came to be seen more as initiation into tertiary study in response to increased participation in university education. As a result, discipline or subject-specific approaches to academic support began to appear. Such approaches range from an academic skills specialist giving a lecture during the regular lecture time on essay writing, based on a real essay prior to its due date; the teaching team and an academic skills specialist drafting week-by-week learning goals and developing a teaching approach to achieve these goals over the semester, addressing any learning problems which might arise; the academic skills specialist writing tutorial lesson plans for subject instructors, after consultation, to achieve both content and academic skills learning goals; and subject coordinators and/or instructors or teaching teams developing their own approach, informed by guidelines offered by academic skills specialists (Cootes, 1994). Experience and research have shown that such context-specific, interactive intervention programs early in a tertiary course can have a lasting positive impact on study strategies, self-directed learning and self-confidence (Chanock, 1994; Cootes, 1994; Garner and Edwards, 1994; Hicks, Irons, and Zeegers, 1994; Johnson and Johnson, 1994; Martin and Ramsden, 1986; Zeegers and Martin, 2001).

A recent example of such research is a longitudinal study at Macquarie University in Australia of first-year students' experience carried out by Krause (2001) in response to Tinto's (1998) identification of the need to integrate first-year students into the university community, combined with Krause's interest in the social nature of the writing process. The study identified two themes within the challenges faced by first-year students: challenges within the broader university context and challenges posed by the research/writing process (Krause, 2001). Krause found that the initial academic writing experience is a "far-reaching and influential vehicle" contributing to the success of the academic integration of first-year students. Based on her findings, she reasons that the integration process in the first year of tertiary study should begin well before the first assignment is due and uses the first writing task as part of this process. Because academic writing at university presents students with new challenges and new conventions, the need for critical thinking and argument development, also often new and to an unknown audience, presents significant pressures. These pressures, often compounded by large class sizes, can be reduced and possibly eliminated if steps to initiate students through the writing process are put in place (Krause, 2001).·

Most universities now recognize that each discipline has its own distinctive discourse, into which many first-year students must be initiated. This is particularly the case in the sciences, with its huge array of subdiscourses. Nowadays, tertiary learning support programs consist of purpose-designed combinations of both the discipline-specific and generic approaches.

The One-to-One Approach

One-to-one tutorial support has long been a standard teaching approach in learning-support units in universities across Australia. There is little doubt among learning support practitioners that individual assistance is the most effective mode of teaching available in that context (Chanock, 1996). It allows for recognition of a student's particular problem and the ability to address it privately on an individual level. It allows us to build students' confidence by recognizing students as individuals and taking them seriously as learners. Often students face the sorts of problems that are not common enough to address in group situations, with the result that they are overlooked or the students experiencing them feel alone and inadequate. For example, learning problems stemming from past learning experiences, special needs, or self-esteem issues. One individual consultation is often enough to overcome the problem, or at least put it into proper perspective, and restore self-confidence in the student. In recent years, however, this mode has been increasingly challenged by questions about its resource efficiency.

The Group Work Approach

Research on how students learn shows that those who work in groups are more successful than those who work in isolation (Kagan cited in Johnston, 2001, p. 171; Trottier, 1999). Student-centered methods of learning, such as group work and discussion, are considered more effective in developing higher-order intellectual skills, such as synthesis and problem solving (Rubin and Herbert, 1998). Also, students who join study groups or are involved in other collaborative learning situations report that they not only learn more but enjoy their work more (Center for Supplemental Instruction, 1992; Goodsell, Maher, and Tinto, 1992). Furthermore, research has shown that an individual student's achievement is positively related to the level of assistance that the student gives to others (Slavin, 1990, cited in Johnston, 2001). According to Tinto (1987), there is a strong correlation between student success and satisfaction and student retention. However, increasing class sizes in tertiary education reduces the scope of application for active and interactive learning techniques.

Student Mentoring

A growing trend in approaches to student learning support across Australia is a preference for student mentoring programs that transcend the boundaries of the approaches outlined previously and combine their best features. Many definitions of student mentoring have been offered, ranging from clearly defined academic advising to broader definitions encompassing emotional/psychological support, professional development, and role modeling (Bond, 1999; Cohen, 1995; Jacobi, 1991). Mentoring can be one to one or involve group work and can be carried out by peer mentors and/or senior mentors.

Student mentoring carried out by peer mentors is becoming increasingly popular for two main reasons. First, it is resource effective. In the current era of increasing class sizes and little hope of increased budgets to improve staffing ratios, it makes good sense to view the student body as a resource. Peer mentoring, also known as peer-assisted learning, incorporates an academic advising role and is viewed by many postsecondary institutions as a powerful influence in promoting learning enrichment (Bond, 1999; Couchman, 1997; Currant, 2001; Jacobi, 1991; Macquarie University Transition and First Year Experience Conference, 2001; Magin and Churches, 1993; University of Missouri–Kansas City, 1995; University of Wisconsin Milwaukee, 2000; University of York, 2002) and an approach that positively affects student learning within targeted courses (Cheah and

Christie, 1996; Kelly, 2000; Sutherland et al., 1996; Worthington et al., 1997; Zeegers and Martin, 1999).

THE FACULTY MENTOR PROGRAM AT UNE

A unique response to the provision of learning support that utilizes the pedagogies outlined previously is a student mentoring program carried out by senior mentors at UNE, known as the Faculty Mentor Program. Funded jointly by a grant held by the Academic Skills Office and each faculty involving the placement of a learning support advisor with a relevant discipline-specific background in each faculty to support first-year, on-campus students. The stated duties of the faculty mentors are to

- liaise with relevant faculty staff to identify learning difficulties specific to discipline areas as experienced by at-risk first-year students;
- interview targeted students to ascertain past and present barriers to academic performance and provide individual advice and guidance;
- advise targeted students on specific courses of action, which might involve accessing currently offered support programs in the Academic Skills Office or UNE's residential colleges (in which a very large proportion of UNE's first-year, on-campus students live), specifically designed short courses, or combinations of both;
- develop and deliver specific learning support short courses for discrete groups of students in faculty groupings; and
- assist academic staff to build these learning support materials into existing course material so that the learning support will be ongoing.

Most of the work of the faculty mentors revolves around such supports as assessment tasks; course-based workshops and short courses on the steps in the research and writing process; working with senior students in the residential colleges to assist students to understand expectations and requirements; collaborating with academic staff to embed academic skills into the curriculum; offering individual advice to students; developing feedback and referral mechanisms for markers; being involved in the resubmit process; providing feedback on marked assignments; and offering assistance to those whose first assignment grades indicate they are at risk.

Although primarily designed as a mentoring program in the academic advising sense, in practice a proportion of the work of the faculty mentors has been of the traditional mentoring kind: supporting, advising, and encouraging students as they encounter and interpret the expectations of university life and study. By taking steps to understand individual student's situations, the program is responding to student need as recommended by Prosser and Trigwell (1999). Early evaluation indicates that this student-centered approach is positively affecting student learning outcomes and easing the first-year transition period for many students.

An investigation into the effectiveness of a brief pilot of this program in 1998, the forerunner to the current program, found that although its effectiveness was not quantifiable, largely because of its short duration (one semester), two factors suggest that the students who sought assistance benefited. First, many students came back repeatedly to the faculty mentor, which suggests that they were deriving value, and second, the mentor received many unsolicited positive comments, most notably after marked assessment tasks were returned (Quinn, Muldoon, and Hollingworth, 2002).

Evaluation of the Faculty Mentor Program

The current program has yielded further qualitative data. After the first year of the current Faculty Mentor Program, sixteen student participants in the program, four from each faculty, were surveyed by phone about their experiences to find out how the program had assisted them in the transition from secondary school to university study. When asked what they learned or gained from their visit(s) to their faculty mentor, the respondents cited assistance with essay writing and referencing, how to be more "academic," reassurance about standards and expectations at university, and effective study techniques. In response to a question about the resulting impact on their studies the students said improved essays, better results in the second semester, improved confidence, lessening of pressure and stress, and the benefits of having received emotional support. All the students surveyed indicated that interaction with their faculty mentor resulted in improved feelings about themselves in relation to university study.

In 2002, midway through the Faculty Mentor Program, a similar survey of forty student participants was carried out by mail. The results showed that the reason that most (thirty-seven of forty) first visited their faculty mentors was to seek assistance about an assessment task and the remainder did so to seek help with study skills in general. More than half of the students surveyed (twenty-five) had attended workshops or short courses offered by their faculty mentors, and all of these either strongly agreed (twenty-two) or agreed (three) that the workshops were useful. All students surveyed strongly agreed (twenty-three) or agreed (seventeen) that their results improved as a consequence of interactions with their faculty mentors. The majority strongly agreed (twenty-two) or agreed (fourteen) that their interactions with their faculty mentors improved their feelings about themselves in relation to university study and their ability to succeed. The majority strongly agreed (twenty-three) or agreed (twelve) that their interactions with their faculty mentors enhanced their ability to complete their first-semester studies. Asked what they learned or gained from their faculty mentor, the majority cited an enhanced understanding of the writing process and improved referencing skills, general study skills, and confidence (see Box 7.1).

Although all four faculty mentors found that the majority of first-year students who sought their assistance had in common issues related to academic writing as their primary area of need, one additional issue stood out as an area of concern to science students. The mentor in the Faculty of the Sciences found that the majority of first-year science students who accessed the program also experienced problems managing their time and organizing their study. She has observed that this appears to stem from two factors related specifically to studying first-year science subjects. First, science students have extremely full timetables and arguably are the busiest students in terms of attendance requirements at lectures, tutorials, and practicals. Second, the fact that many have not studied specialized science subjects at school (as described by Quinn in Chapter 10) means that their workload is further increased by the need to catch up in disciplines such as chemistry and mathematics, which are either compulsory subjects or prerequisites for other subjects. A lack of background knowledge and organizational skills seem to go hand in hand. For example, many

---

**BOX 7.1. Selected Comments from Students
on the Faculty Mentor Program
About Improved Confidence**

- [I gained] more confidence, particularly about writing; the confidence I needed; confidence to kick on; the feeling that I could actually do it; a lot—it made me feel more positive about my work; confidence in completing required tasks, and learned not to get so worried; faith in my abilities in referencing and essay writing [after visiting the faculty mentor].
- [I felt] more confident about surviving; more relaxed and reassured; very positive; able to talk to other lecturers without fear; on the right track [after visiting the faculty mentor].
- [I achieved] improved written expression; improved grammar; proofreading skills, [the ability to] write a good, structured, logical, emotion-free essay; better results in Semester 2; better marks; better results; much better results—got a HD, which wouldn't have happened otherwise; good results—went from Ps to Ds; added polish to essays [after visiting the faculty mentor].
- I wouldn't have survived the year otherwise!

---

chemistry students who have not previously studied chemistry often need assistance with simply getting started. This problem is exacerbated by an inability to appreciate the relevance of these subjects in the first year. At UNE, where the majority of first-year on-campus students also live in the residential colleges, these time-management issues are compounded by the social, cultural, and sporting distractions inherent in the residential experience.

Summary and Implications

The Faculty Mentor Program reaps the benefits to be derived from discipline-specific learning support as well as the advantages of both group work and one-to-one learning support. It appears to be especially valuable to those who do not have the skills and background knowledge once considered essential to university study by offering an element of socialization into these requirements and expectations.

According to Tinto (1993), "academic difficulty" is one of the most common causes of attrition, and this is confirmed by research carried out at UNE (Centre for Higher Education Management and Policy, 1999) and other Australian studies (Johnston, 2001; Krause, 1998). Part of the problem for many students is having to adjust preconceived expectations and become accustomed to different requirements and approaches to assessment. In the sciences, this can be particularly critical because of the often bewildering array of disciplines and conventions. The Faculty Mentor Program uses early assessment tasks as a vehicle to address these issues and assist in the process of academic integration (see Box 7.2).

**BOX 7.2. Selected Comments from Students
on the Faculty Mentor Program
About Improved Learning Outcomes**

- [I gained] the fundamentals of good study habits and scientific writing that one tends to forget after an absence from study; a better grasp of what is required at uni. I used to be a mechanic so had no idea at all; reassurance re: standards; an understanding of what the science faculty wanted; good tips about studying; effective study techniques; useful advice; academic skills information that was clear and easy to understand [after visiting the faculty mentor].
- [I learned] how to be more academic; what the lecturers are expecting.
- I have a different perspective toward my studies after contacting the faculty mentor. The mentor has greatly motivated me to study, and with a new, clearer view of what I am doing than before.
- The sciences faculty mentor really made the questions make sense.
- I felt very comfortable with the sciences faculty mentor. She did not speak down to me but at my level and went out of her way many times to help me.
- It was only during the first weeks of semester one that I visited the sciences faculty mentor. It mainly concerned my ability to start off with CHEM110. She really helped me.
- Without her [I] would have dropped out.
- When you don't know where to start or what the question is really asking, the faculty mentor can really reassure you and help.
- She pointed me in the right direction and made me think about what needs to be done.

The Faculty Mentor Program also built bridges between all the stakeholders in the first-year experience—the faculties, the residential colleges, the student support services, and, most important, the first-year students. It promoted interaction in the learning process, which not only enhances the quality of learning but also contributes to students' sense of belonging within the learning community and their sense of competency. Evidence gained from evaluation of the program shows that it strengthened students' sense of connectedness to the institution (see Box 7.3). The importance of this in terms of student retention and the value of it occurring very early in the transition process is huge (Nora, 1993, cited in Krause, 2001; Levin and Levin, 1991; National Resource Centre for the First Year Experience and Students in Transition, 2001; Tinto, 1993).

However, not all reasons for student withdrawal are about educational issues (Pitkethly and Prosser, 2001), and this is also supported by a recent survey carried out at UNE, which indicates a range of other 'life' issues affecting students' ability to settle into and persevere with their studies. These include financial worries, (paid) work pressure, time-management difficulties, family and relationship problems, and low self-esteem (University of New England, 2002). The faculty mentor clearly plays a role in this regard as well as in terms of

---

BOX 7.3. Comments from Academic Staff in the Faculty of the Sciences, UNE

- "It is my opinion that this is an extremely valuable component of the faculty's offerings to students, acting as a bridge between academia and the students." (lecturer in the School of Biological, Biomedical, and Molecular Sciences)
- "The communication link between lecturer and student is invaluable." (first-year biology coordinator)
- "Faculty mentor has made an important contribution to our first-year students. I have had many positive comments back from students. Because she is available, it has reduced the workload of staff in marking—we can refer students to her for assistance in how to approach assignments, writing techniques, time management, and so on. I consistently refer students to (the Sciences Faculty Mentor) for help in how to take exams and how to prepare for them." (associate dean, Teaching and Learning, in the Faculty of The Sciences)

offering emotional support and referral to other student support services, such as the Counseling and Careers Service (see Box 7.4).

CONCLUSION

If we are to improve student learning outcomes and retention rates in the sciences within existing resource constraints it is clear that what is needed is a support network that utilizes the known benefits of context-specific learning support while also aiming to achieve academic and social integration. This chapter has provided an overview of current approaches to learning support and described one successful approach in the Faculty of the Sciences at UNE. The mentoring program described is underpinned by the literature on student retention while exemplifying best practice in student learning support. It combines the best features of discipline-specific academic skills development and learning support with an appropriate balance of one-to-one counseling and group work. Mentoring is a student-centered approach that runs parallel with, complements, and counterbalances traditional teaching methods for the benefit for all stakeholders, particularly students.

**BOX 7.4. Selected Comments from Students
on the Faculty Mentor Program
About Emotional Support**

- [The faculty mentor] took the pressure off—if in doubt I knew she would help; gave me lots of tips and emotional support, even "life skills"; was really useful for me especially at the beginning of my first semester, to adapt to university life and with the work load; provided a good sounding board; helped get past "brick wall"; helped me get a good, well-balanced routine going; made my transition from year twelve to UNE smoother by her encouragement and services she provided.
- I have learned how to balance my studies, sport, and social life. Consequently, I have enjoyed my time at uni while maintaining positive results in my courses thus far.
- The faculty mentor taught me how to enjoy [uni] by keeping in control of the workload as well as other commitments.
- Knowing the "safety net" was there was invaluable. It settled my mind to be able to pop in for a talk.

REFERENCES

Baldauf, R.B. Jr. (1997). Tertiary language, literacy and communication policies: Needs and practice. Paper presented at the First National Conference on Tertiary Literacy: Research and Practice, Victoria University of Technology, Melbourne, Australia, pp. 1-19.

Beasley, C. (1997). Students as teachers: The benefits of peer tutoring. Paper presented at the Learning Through Teaching, 6th Annual Teaching Learning Forum, Murdoch University, Australia. February.

Bond, A. (1999). *Student mentoring: Promoting high achievement and low attrition in education and training.* Leabrook, Australia: National Centre for Vocational Educational Research (NCVER).

Campus Review (2003). Survey points concern at 40 percent drop-out figure. *Campus Review,* March 12-18: 2.

Center for Supplemental Instruction (1992). *Supplemental instruction: Theoretical framework, review of research concerning the effectiveness of SI from the University of Missouri–Kansas City and other institutions from across the United States.* Kansas City: Center for Academic Development, University of Missouri.

Centre for Higher Education Management and Policy (1999). *Student intention and progression survey, stage 2.* Armidale: University of New England, Australia.

Chanock, K. (1994). Disciplinary subcultures and the teaching of academic writing. Paper presented at Integrating the Teaching of Academic Discourse into Courses in the Disciplines, La Trobe University, Australia, November 21-22.

Chanock, K. (1996). The "interdiscourse" of essays: Listening one-to-one and telling one-to-one hundred. Paper presented at What Do We Learn from Teaching One-To-One That Informs Our Work with Large Numbers? La Trobe University, Australia, November 18-19.

Cheah, S. and Christie, R. (1996). Mentoring in the Faculty of Information Technology 1995. Paper presented at Transition to Active Learning. Second Pacific Rim Conference on the First Year in Higher Education, University of Melbourne, Australia, July 3-5.

Cohen, N. (1995). *Mentoring adult learners: A guide for educators and trainers.* Malabar, Florida: Krieger Publishing Company.

Cootes, S. (1994). Issues in collaboration between academic skills teachers and subject teachers: Two approaches to integrating academic skills teaching with subject tutorials. Paper presented at the Integrating the Teaching of Academic Discourse into Courses in the Disciplines, La Trobe University, Australia, November 21-22.

Couchman, J. (1997). *Supplemental instruction: Peer mentoring and student productivity, "researching education in new times."* Brisbane: Office of Preparatory Studies and Continuing Studies, University of Queensland, Australia.

Currant, B. (2001). *Stepping stones to success: Make the most of your Bradford experience.* University of Bradford, Bradford, England.

Department of Education Training and Youth Affairs (2000). *Students 1999: Selected higher education statistics.* Canberra, Australia: Commonwealth Government.

Garner, M. (1997). Some questions about integrated communication skills programmes. Paper presented at the First National Conference on Tertiary Literacy: Research and Practice, Melbourne, Australia, March 14-16, 1996.

Garner, M. and Edwards, H. (1994). Integrating academic discourse—What can we learn from experience? Paper presented at Integrating the Teaching of Academic Discourse into Courses in the Disciplines, La Trobe University, Australia, November 21-22.

Goodsell, A., Maher, M., and Tinto, V. (1992). *Collaborative learning: A sourcebook for higher education*. University Park, PA: National Center on Postsecondary Teaching, Learning, and Assessment.

Hicks, M., Irons, E., and Zeegers, P. (1994). Academic and communication skills taught in science and engineering courses. Paper presented at Integrating the Teaching of Academic Discourse into Courses in the Disciplines, La Trobe University, Australia, November 21-22.

Jacobi, M. (1991). Mentoring and undergraduate academic success: A literature review. *Review of Educational Research,* 61(4): 505-532.

Johnson, D. and Johnson, R. (1994). *Learning together and alone: Cooperation, competition, and individualization* (Fourth edition). Boston: Allyn and Bacon.

Johnston, C. (2001). Student perceptions of learning in first year of economics and commerce faculty. *Higher Education Research and Development,* 20(2): 169-184.

Kelly, B. (2000). *Faculty based peer assisted study sessions (PASS) at University of Queensland.* Brisbane: Teaching and Educational Development Unit, University of Queensland, Australia.

Krause, K. (1998). Writing assignments in the first year: Student perceptions and strategies for survival. Paper presented at the Third Pacific Rim Conference: First Year in Higher Education—Strategies for Success in Transition Years, July 5-8, Auckland, New Zealand.

Krause, K. (2001). The university essay writing experience: A pathway for academic integration during transition. *Higher Education Research and Development,* 20(2): 147-168.

Laws, P.M. (1996). Undergraduate science education: A review of research. *Studies in Science Education,* 28: 1-85.

Levin, M. and Levin, J. (1991). A critical examination of academic retention programs for at-risk minority college students. *Journal of College Student Development,* 32(4): 323-334.

Macquarie University Transition and First Year Experience Conference (2001). Macquarie University, Sydney, Australia.

Magin, D. and Churches, A. (1993). Student proctoring: Who learns what? Paper presented at the Peer Tutoring: Learning by Teaching Conference, University of Auckland, New Zealand.

Martin, E. and Ramsden, P. (1986). Do learning skills courses improve student learning? In J. Bowden (Ed.), *Student learning: Research into practice* (pp. 149-166). Parkville, Australia: University of Melbourne.

McInnes, C. (2001). Researching the first-year experience: Where to from here? *Higher Education Research and Development,* 20(2): 105-114.

McInnes, C. and James, R. (1995). *First year on campus: Diversity in the initial experiences of Australian undergraduates*. Melbourne: Australian Government Publishing Service.

McInnes, C., James, R., and Hartley, R. (2000). *Trends in the first-year experience.* Canberra: Australian Government Publishing Service.

National Resource Centre for the First Year Experience and Students in Transition (2001). FYE courses increase retention among at-risk students. *FYE,* 14(1): 10-11.

Niland, J. (1998). The fate of Australian science—The future of Australian universities. Address to the National Press Club, February 25. Australian Vice-Chancellors' Committee. Retrieved March 18, 2003, from the World Wide Web: <www. avcc.edu.au/news/public_statements/speeches/1998/fate.htm>.

Parker, L.H. (1997). Institutional practices in promoting tertiary literacy: The development and implementation of a university-wide policy for enhancing students' communication skills. Paper presented at the First National Conference on Tertiary Literacy: Research and Practice, Melbourne, March 14-16.

Pitkethly, A. and Prosser, M. (2001). The First Year Experience Project: A model for university-wide change. *Higher Education Research and Development,* 20(2): 185-198.

Prosser, M. and Trigwell, K. (1999). Relational perspectives on higher education teaching and learning in the sciences. *Studies in Science Education,* 33: 31-60.

Quinn, F., Muldoon, R., and Hollingworth, A. (2002). Formal academic mentoring: A pilot scheme for first year students at a regional university. *Mentoring and Tutoring,* 10(1): 21-35.

Rubin, L. and Herbert, C. (1998). Model for active learning: Collaborative peer teaching. *College Teaching,* 46(1): 26-31.

Solomon, J. and Thomas, T. (1999). Science education for public understanding of science. *Studies in Higher Education,* 33: 61-89.

Sutherland, L., Ingleton, C., Cowie, J., and Marshall, V. (1996). Using peer group tutorials to enhance learning in a repeat first year macroeconomics class. Paper presented at the 25th Conference of Economists, University of Adelaide, Adelaide, Australia, September 24-27, 1995.

Taylor, P. (2000). Preparing students for successful participation in new learning environments. Paper presented at Flexible Learning for a Flexible Society, Toowoomba, Australia. July 2-5.

Threadgold, T., Absalom, D., and Golebiowski, Z. (1997). Tertiary literacy conference summary: What will count as tertiary literacy in the Year 2000? Paper presented at the First National Conference on Tertiary Literacy: Research and Practice, Melbourne, Australia, March 14-16, 1996.

Tinto, V. (1987). *Leaving college: Rethinking the causes and cures of student attrition.* Chicago: University of Chicago Press.

Tinto, V. (1993). *Leaving college: Rethinking the causes and cures of student attrition* (Second edition). Chicago: University of Chicago Press.

Tinto, V. (1998). Colleges as communities: Taking research on student persistence seriously. *The Review of Higher Education,* 21(2): 167-177.

Trottier, R. (1999). A peer-assisted learning system ("PALS") approach to teaching basic sciences: A model developed in basic medical pharmacology instruction. *Medical Teacher,* 21(1): 43-47.

University of Missouri–Kansas City (1995). *Supplemental instruction supervisor manual.* Kansas City: University of Missouri.

University of New England (2002). *2002 First Year Experience Survey.* Armidale: Admissions Committee, University of New England, Australia.

University of Wisconsin Milwaukee (2000). Peer Mentoring Centre. November 30. College of Letters and Science. Retrieved June 17, 2002, from the World Wide Web: <http://www.uwm.edu/letsci/edison/pmc/index.html>.

University of York (2002). The York Award: Skills for employment and life. University of York. Retrieved October 18, 2002, from the World Wide Web: <http://www.york.ac.uk/admin/ya/>.

Worthington, A., Hansen, J., Nightingale, J., and Vine, K. (1997). Supplemental instruction in introductory economics: An evaluation of The University of New England's peer assisted study scheme (PASS). *Australian Economic Papers* (Special Issue): 69-80.

Yorke, M. (2001). Formative assessment and its relevance to assessment. *Higher Education Research and Development,* 20(2): 115-126.

Zeegers, P. and Martin, L. (2001). A learning-to-learn program in a first-year chemistry class. *Higher Education, Research, and Development* 20(1): 35-52.

Chapter 8

"Drowning by Numbers":
The Effectiveness of Learner-Centered
Approaches to Teaching Biostatistics
in the Environmental Life Sciences

Debra L. Panizzon
Andrew J. Boulton

Learning is not viewed as merely a cumulative accretion of knowledge by a largely passive learner, but as an active process in which the learner is engaged in constructing or generating concepts to account for novel phenomena. (Cleminson, 1990, p. 439)

INTRODUCTION

Many environmental science students at the tertiary level struggle with topics that entail biostatistics and applied mathematics, yet virtually all life sciences have quantitative aspects that demand statistical evidence to support or refute predictions in the hypothesis-based scientific paradigm (Elzingha et al., 2001). Increasingly, environmental scientists are required to statistically demonstrate the effectiveness of resource management programs. They must be able to establish that their survey designs and resultant data are sufficiently

Some of this research was done as part of the thesis requirements for a graduate certificate in higher education by AB, supervised by Associate Professor C. McLoughlin. We thank the students who took EM331 from 1997 to 2002, especially those who allowed us to interview them in 2000.

141

powerful in a statistical sense (e.g., Fairweather, 1991) to withstand scientific and legal scrutiny in environmental issues that may involve protection of World Heritage–listed national parks or natural resource exploitation worth millions of dollars. In short, life scientists need mathematics as much as do physicists and engineers but seldom do such students show much natural predilection for numbers. So what approaches can teachers use that might overcome the fear of "drowning by numbers"? How can we help students conquer this fear and start to learn effectively? What does educational theory have to offer teachers tackling this challenge at a tertiary level? How do learner-centered approaches lend themselves to encouraging a deep approach to learning about biostatistics in the life sciences?

Over the past few decades, research in teaching and learning in science has shown that students enter formal education with well-developed ideas about scientific concepts that have been formulated predominantly through sensory experiences with their natural world. Although these conceptions are coherent and make sense to the student, they may differ significantly from the scientific view (Driver et al., 1994). To enable students to bridge the gap between their intuitive conceptions and the contemporary scientific view, there must be a significant change in conceptual knowledge. As a result, the role of the student changes from that of a *passive recipient* to an *active constructor* of knowledge. This new role for the learner, as intimated in the opening quote, has been engendered by the constructivist approach prevalent in current science and mathematics teaching and learning (Matthews, 1993). Its fundamental premise is that learning involves the construction of knowledge by individuals using their existing conceptions. The recognition of the importance of these pre-existing ideas in the learning process has led to an explosion of research studies in the past two decades exploring the nature and complexity of the scientific conceptions held by students (summarized in Pfundt and Duit, 1994).

In this chapter, we selectively review current educational theories about the way students learn, with specific applications in the life sciences and biostatistics. We commence with an exploration of constructivism, highlighting the importance of social engagement and educational context. Following this, we explore the issue of alternative conceptions as a means of explaining why students retain incorrect views about scientific concepts even after formal instruction. We outline a number of teaching strategies and learning opportunities provided to help students undertaking a third-year course in survey design and biostatistics (EM331) to construct a deeper understanding of statistics. The contrast between approaches to learning is also explored because, for many of these students, inappropriate learning strategies such as repetitive reading of texts failed to help them understand the theory underlying the application of statistics and how to incorporate it into their cognitive framework. Using results from questionnaires and interviews conducted with tertiary environmental science students, we assess the effectiveness of teaching strategies that demonstrate explicitly the relevance of the learned material to students' career aspirations, as well as activities that encourage student social engagement with the material, and we specifically address potential sources for alternative conceptions. We conclude that the integration of constructivist theory with an understanding of alternative conceptions and ways of resolving them creates a powerful tool for teachers seeking to help students in the life sciences adopt a "deep approach" to learning biostatistics and survey design.

THE CONSTRUCTIVIST FRAMEWORK: TWO COMMON THREADS

There is a long history of constructivist views in the philosophy and practice of disciplines as disparate as education, developmental psychology, human sexuality, and computer technology (Noddings, 1990). Regardless of the area of study, the major premise of constructivism is that individuals construct their own meaning or "versions of reality" (Sutherland, 1992, p. 79) from their own experiences. In this context, learning is no longer viewed in terms of knowledge acquisition by a passive learner, but as an active process requiring the learner to construct new ideas or restructure existing ones (Bell, 1993; Costa, Hughes, and Pinch, 1998). To promote this type of learning, Watts (1994) recognized the need for a "good constructivist classroom" (p. 52). He proposed that such a classroom should create an environment in which individuals are encouraged to take responsibility for their own learning by setting specific goals within the confines of their abilities and working purposively toward these self-determined ends. In environmental science, the classroom must be extended to the natural world, and it is the need to describe the variance and extraordinary complexity of this natural world that first assails the quantitative life scientist.

Although this proposition of constructivism is shared by many educationalists, Good (1993) suggests that multiple versions of constructivism actually exist. In support of this view, Geelan (1997) identified a number of broad categories of constructivism (Table 8.1). Personal constructivism describes the active acquisition of knowledge as an individual, whereas social constructivism emphasizes the "group approach" and the need for interactions among learners (Ernest, 1992, p. 1003). Contextual constructivism extends this further to incorporate other aspects, such as culture, economics, and educational level, that help create different learning contexts. This is the reason we need to change our teaching approaches when working with bachelor of science as opposed to bachelor of rural science stu-

TABLE 8.1. Summary of the Major Forms of Constructivism and Their Characteristics

Form of constructivism	Characteristics
Personal constructivism (Kelly, 1955; Piaget, 1972; Driver and Oldham, 1986)	Knowledge is actively built up by individuals based on their prior conceptions and experiences and is not transferred from one situation to another.
Social constructivism (Solomon, 1987; Tobin, 1990; Ernest, 1992, 1993)	Knowledge is not constructed in isolation of others but occurs as a result of social interactions. For example, understanding language emerges from interactions with family and the wider community. Similarly, the knowledge of science has been socially constructed by consensual agreement among the scientific community or culture and should not be ignored.
Contextual constructivism (Cobern, 1993)	Knowledge is constructed not just in relation to social interactions but is influenced by the learning context. This could include factors such as culture, occupation, geographic location, economics, or educational levels. As a consequence of these factors, science is often not relevant to many cultural groups existing within Western nations. In the United States, for example, women, African Americans, and Hispanics are still underrepresented in the sciences (Gallard, 1993).

dents, for example. These versions of constructivism highlight two crucial features that must be considered in any teaching situation. The first is the value of social engagement to enhance the construction of meaning and understanding by the learner. In life sciences and natural resource management, this social engagement is de rigueur—not only are scientists likely to work with each other because the magnitude of most environmental issues requires a multidisciplinary approach, but the social aspects of communication of these issues to the general public, politicians, and resource managers are just as important. We return to the subject of social engagement later.

The second crucial feature is the panoply of complex variables that comprises the background and experiences of the learner which are brought to any learning situation. Although these factors are often given consideration in primary and secondary learning environments, they are usually overlooked at the tertiary level. Yet students bring their cultures, beliefs, previous experiences, and conceptions with them when they enter the lecture theater or laboratory. In the case of biostatistics, these previous experiences may be negative and even disabling. Over the past five years, consecutive classes of twenty-five to thirty-five environmental science students about to undertake a tertiary course in survey design and biostatistics (EM331) have been asked to outline their previous experiences in biostatistics. Responses are remarkably consistent (see Box 8.1) and indicate a fear of numbers, uncertainty about which test to use and when (regardless of the question), and confusion about formulae, symbols, and applications. Prompted by this consistency in response, we explored the effectiveness of learner-centered approaches rather than classical lecturer-student pedagogy (the norm for students prior to taking EM331) in enhancing students' grasp of the fundamentals of survey design and data analysis.

CHALLENGING ALTERNATIVE CONCEPTIONS IN BIOSTATISTICS

Within science, there is often a substantial difference between the conceptions held by students and those held by the scientific community (Osborne and Wittrock, 1983). In many cases, the conceptions of the student may be "not quite correct." Within the constructivist

---

**BOX 8.1. Summary of Student Responses
to the Question "Why They Hate
or Fear Classes in Statistics"**

"I hate/fear classes in statistics because . . ."

1. Too much use of complex formulae, Greek letters, equations with letters
2. Little relevance to environmental issues/limited relevance to other units being done in classes
3. Not clear about which statistic to use to answer which question. How do I know what statistic to use?
4. Fear of numbers. ("I did biology to avoid mathematics and now I find I need to learn statistics.")
5. Not enough time to learn the ideas; covered too quickly and too confusing

Note: These are the top five responses, arranged in order of first mention, by groups of students in EM331 from 1997-2001.

---

framework, this is termed an *alternative conception* or *misconception*. The role of the teacher is to help bridge this gap by changing the alternative conception so that the student acquires an understanding of the models and concepts of conventional science. However, research undertaken in this area over the past few years highlights the tenacity of these alternative conceptions (Driver et al., 1994). It is one thing to recognize that your students possess them, but trying to remedy the situation is another altogether. Couple this with negative prior experiences in statistical science, and the challenge is clear.

The reason that conceptions, whether correct or not, are tenacious is that the student has developed the concepts and schema over a long period of time. A schema is a framework constructed by an individual comprising concepts, knowledge, relationships between concepts, and past experiences. It is stored in the long-term memory and influences how an individual interprets new information (Pines and West, 1983). Howard (1987) identified four possible reasons as to why the alternative conceptions held by students persist, each relating to the schemata (pl) held by the students. We reviewed these and put them in a real context, using data obtained from students enrolled in our biostatistics course. The data included observations made during classes, questionnaires undertaken by all the students, and interviews conducted with eight of the students.

1. The schemata are still evolving because concepts are still being developed within the individual. Concept acquisition is a time-consuming process involving the incorporation of new knowledge along with the possible reorganization of the existing schemata (Bell and Freyberg, 1985). Therefore, the student's inability to understand a new concept may result from the premature introduction of the concept before the existing cognitive structure has sufficiently developed. This is truer in biostatistics than in many other fields of life science because knowledge and understanding of aspects of statistics must build on prior understanding. For example, we tested the assumption that students who had done first-year statistics fully understood the mathematical principles underlying expression of statistical variance and confidence intervals. Student responses to a "background knowledge probe" conducted at the start of EM331 revealed that only fourteen of thirty-five students could define a confidence interval, and of those a mere nine could apply the formula correctly. But all thirty-five students had "learned" about confidence intervals in a first-year statistics course. Most students knew confidence intervals "had something to do with variance" or "variability about the mean" (EM331 student 2). However, when they were asked a related question about how to express variance about the mean of a sample, they did not mention confidence intervals, standard errors, or standard deviations. Lack of opportunity to apply this material covered in first year may slow the rate of evolution of the relevant schema, and at least one life science curriculum being reviewed at the University of New England explicitly recognizes the importance of sequential development of knowledge construction in biostatistics.

2. The sufficiency of existing schemata identifies the possible inadequacy of the established conceptual structures for dealing with new information. Problems in this regard arise when students approach tertiary-taught science concepts with the perception that the context they encounter has little intrinsic value. As a result, some students are not prepared to "go to the great trouble of trying to understand them" (Howard, 1987, p. 190), and so rely on their existing schemata. In many cases, these schemata are insufficient,

given the students' limited range of experiences compared to those of scientists. Consequently, as students are unaware of the array of phenomena that need to be taken into account by the schemata, they rely heavily on their existing structures, resulting in an alternative understanding of the scientific event (Osborne and Wittrock, 1985). The students' responses as to why they "hate statistics" (Box 8.1) partly hints at this view. When asked to elaborate on their answers, a common response was that "we probably won't need to know much of this stuff in future" (EM331 student 4)—a surprising reply given that they had opted for an elective course that was going to include statistics. There was also a common retort that "the stuff (statistics and formulae) is in books or on the computer so we don't need to understand it" (EM331 student 3). This indicated some confusion between *accessing* the material versus *understanding* when to use particular statistical tests or formulae. Reassuringly, responses to interviews with students after they had taken the course seemed to indicate that they had grown to appreciate the relevance of the material that was covered and grasped the central theme of knowing which statistical test to use rather than worrying about rote learning formulae. For example,

> Every course I have been doing since, I have been trying to use statistics in it, so my knowledge of statistics is increasing whereas before completing EM331 it was quite shocking. Could hardly do anything! Now, I not only grasp the statistics but feel really confident in being able to select the most appropriate statistical method for analyzing my data. (EM331 student 1)

3. The pressure of work and time are two factors that have the potential to inhibit the development of schemata. Students are confronted with many subjects at any one time, with a large number of new concepts associated with each of the subjects. As they are seldom given adequate time to understand all the concepts necessary within each subject, many students become dependent on "survival" methods, such as rote learning or strategically allocating effort according to the assessment scheme. Biggs (1989) refers to these as surface learning or "achieving" learning approaches. There is considerable evidence that such pressures exist during the first year of study when students are undertaking introductory statistics courses, and a common perception appears to be that early univer-

sity courses are broad, shallow, and general, whereas they "don't get interesting or relevant" (EM331 student 5) until the final year of study. Nonetheless, a related study exploring "deep" and "shallow" approaches to learning reveals that even in this third-year level course, only eight of twenty-six students actually demonstrated possession of traits characteristic of a deep approach to learning based on Biggs' (1989) criteria.

4. Some new schemata are difficult to accept because they differ markedly from existing structures. The classic example is the difficulty experienced by some students in accepting Darwin's theory of evolution because it conflicts with their existing religious beliefs or schemata. Ausubel (1977), however, believed that the acceptance of new schema failed because of a lack of available cognitive structure within the learner. As a consequence, the student cannot anchor new ideas to any existing cognitive scaffolding and may need to rely on rote learning. Ultimately, this lack of anchorage results in the new material remaining unstable or ambiguous in meaning, and it is retained for only short periods of time. Plausibly, this also accounts for the apparently poor retention of material learned in first-year statistics courses.

ENCOURAGING A DEEP APPROACH TO LEARNING

Once a teacher is able to identify the alternative conceptions and how they may arise, progress can be made toward changing these by trying to establish good learning habits that encourage a deep approach to learning. We have adopted several teaching strategies in EM331 that aim to address the issues previously identified. These include trying to

- enhance students' social engagement with the material;
- demonstrate the relevance of biostatistics in the life sciences;
- address negative perceptions about biostatistics;
- identify the conceptions held by students prior to the introduction of a new unit so as to either build upon them or help students restructure them; and
- adopt assessment methods that favor active learning, timely reflection, and specific responses to consecutive tasks whose completion provide the foundations for the next one.

Learner-centered approaches have been demonstrated to be highly effective at promoting a "deep-level" understanding, largely because of shifts in student motivation toward intrinsic desires to satisfy their curiosity (e.g., Marton and Saljo, 1976; Marton, Dall'Alba, and Beaty, 1993); development of a feeling of "ownership" of their learning product and a desire to interact with other students (collaborative learning, Biggs, 1989); and extensions of learning experiences to other situations (Crawford et al., 1998). Building on this literature and given the prevalence of alternative conceptions in biostatistics, we have tried and tested several exercises in EM331, taking care to match assessment exercises with our learning objectives (Toohey, 1999).

The first broad perception to overcome was the view that biostatistics and survey design was "too hard to understand" and of limited value in real life, hence not worth the effort. We addressed this by asking students to nominate an issue that they *did* feel was important and describe how they might design a research survey or experiment to answer some hypotheses about the issue. Surprisingly, in five years of running this course, we have never had a student who has complained that he or she is unable to think of an important environmental issue for which he or she cannot design a relevant survey or experiment. By beginning with the students' identified areas of interest and then demonstrating how statistical methods can be applied to these areas, we are helping students build their knowledge within a specific context. Although the students had undertaken a statistics course in their first year, our strategy was to provide the context and then incorporate the most appropriate statistical tests to interpret the data. Students' responses to the strategy have been very positive:

> The fact that people were talking about their own areas of interest and possible research projects for the course made the statistics incredibly relevant. Coming out of this part of the course I had a much better idea about the number of samples needed and the type of analysis most appropriate for the data collected. (EM331 student 3)

We follow up this relevance issue by explaining to students that a valuable graduate skill in life sciences is the ability to prepare a grant application seeking funding for a research project, as well as the expertise to assess grant applications to ensure they are feasible, economical, and well designed. Subsequently, one of the assessment items of EM331 entails the preparation of a grant application using an electronic application form modeled on the format used for internal university grants. This includes some effort to generate a budget of time and money, describe the significance of the research, specify in detail the design and data analysis, and explain the benefits of the findings.

To enhance social engagement with the material and give students experience in grant proposal assessments, we organize students into panels of peers who address similar environmental issues. For example, all students who are doing bird surveys as the basis of their project would form a bird panel and, during a three-hour practical session, review and mark their peers' proposals using the assessment sheet that will ultimately be used by the lecturers in the course. Students submit their final draft of the proposal after they have had a week to act upon the suggestions of their peers, fully acknowledging the suggestions and comments. Students interviewed about this process were unanimous in their praise of the value of peer assessment, and several students admitted that they continued discussion with their peers during the project, sharing ideas, literature, and sometimes even doing fieldwork together. For example,

> It worried me a bit at first but when we came down to assessing one another, everyone took it seriously and really tried to offer constructive comments. I learned so much by hearing what my peers working in similar areas were thinking. For me it also initiated discussion with people doing similar projects—people I would not normally spend much time talking to. I now realize the importance of chatting with people about your work and how much can be gained from the shared discussion. Even when we got to doing our individual research projects, we helped one another out in the field. (EM331 student 2)

After the students had developed their own research proposal, assessed several of their colleagues' proposals, and had comments from the lecturers who had marked the proposals and assessments, a robust design with clear-cut field and analytical methods emerged. For further assessment, the students conducted the surveys or experiments and then presented a seminar on their findings, again to a panel of peers who assessed the work and presentation of results. This consolidated the learning experience further, as well as enabling assessment of a different form of communication and, for some students, this was their first exposure to oral presentation of their own work. Peer questions and comments were usually suitably compassionate and, even at this late stage in the course, alternative conceptions (usually regarding statistical analysis) were still being revealed.

CONCLUSION

Understanding the ways in which knowledge is acquired and constructed by our students is crucial if we are to address their needs at the tertiary level. If learning is to be enhanced, we need to think carefully about what prior knowledge and alternative conceptions are possessed by our students, and then provide a range of teaching strategies to support or contradict this knowledge so as to promote reconstruction by the students. In some topics, such as biostatistics in the life sciences, the challenge is exacerbated by negative past experiences sometimes bordering on fear and denial.

In this chapter, we have presented several teaching approaches that have evoked deep-level approaches to learning by leading students to acknowledge the relevance of biostatistics in many branches of the life sciences and, more importantly, apply these to field situations in a hands-on approach. The sequence of research proposal, peer assessment, project data collection, and analysis, seminar, and final report presentation meant that students continually built on prior experiences, explicitly linked to the assessment of the course. By creating real-life scenarios (applying for a research grant, assessing peers' grant proposals, conducting environmental research, and analyzing real data) in the classroom, students learn by doing and develop a deeper understanding of the application of biostatistics in the environmental sciences. Interactions with their peers and exposure to broader contexts (both social and environmental) enrich the learning experience and consolidate the understanding of this material.

Throughout the whole sequential course, the learner has been at the center of the experience—a stark contrast to the more usual presentation of biostatistics as exposure of students to sets of worked examples, lectures on a variety of statistical methods, and case studies based on other people's work. Our closing message for teachers in this field is to encourage them to put their learning exercises into a context that allows students more ownership of the project and the data, provides continual feedback to the student during the development of the project (proposal, peer review, project seminar, final report), and switches the emphasis from a teacher-centered model, where examples are presented for students to work through, to a learner-centered model, where students savor the experience themselves.

REFERENCES

Ausubel, D.P. (1977). Cognitive structure and transfer. In N. Entwistle and D. Housell (Eds.), *How students learn* (pp. 93-104). Lancaster, UK: Institute for Research and Development in Post Compulsory Education at University of Lancaster.

Bell, B. (1993). *Children's science, constructivism, and learning in science.* Deakin, Australia: Deakin University Press.

Bell, B. and Freyberg, P. (1985). Language in the science classroom. In R. Osborne and P. Freyberg (Eds.), *Learning in science* (pp. 29-40). Auckland, New Zealand: Heinemann Press.

Biggs, J. (1989). Approaches to the enhancement of tertiary teaching. *Higher Education Research and Development,* 8: 7-25.

Cleminson, A. (1990). Establishing an epistemological base for science teaching in the light of contemporary notions of the nature of science and how children learn science. *Journal of Research in Science Teaching,* 27(5): 429-445.

Cobern, W.W. (1993). Contextual constructivism: The impact of culture on the learning and teaching of science. In K. Tobin (Ed.), *The practice of constructivism in science education* (pp. 51-70). Hillsdale, NJ: Lawrence Erlbaum Associates.

Costa, S., Hughes, T.B., and Pinch, T. (1998). Bringing it all back home: Some implications of recent science and technology studies for the classroom science teacher. *Research in Science Education,* 28(1): 9-21.

Crawford, K., Gordon, S., Nicholas, J., and Prosser, M. (1998). Qualitatively different experiences of learning mathematics at university. *Learning and Instruction,* 8: 455-468.

Driver, R. and Oldham, V. (1986). A constructivist approach to curriculum development in science. *Studies in Science Education,* 13: 105-122.

Driver, R., Squires, A., Rushworth, P., and Wood-Robinson, V. (1994). *Making sense of secondary science: Research into children's ideas.* London, UK: Routledge.

Elzingha, C.L., Salzer, D.W., Willoughby, J.W., and Gibbs, J.P. (2001). *Monitoring plant and animal populations.* New York: Blackwell Science.

Ernest, P. (1992). The nature of mathematics: Toward a social constructivist account. *Science and Education,* 1: 89-100.

Ernest, P. (1993). Constructivism, the psychology of learning, and the nature of mathematics: Some critical issues. *Science and Education*, 2: 87-93.

Fairweather, P.G. (1991). Statistical power and design requirements for environmental monitoring. *Australian Journal of Marine and Freshwater Research*, 42: 555-567.

Gallard, A.J. (1993). Learning science in multicultural environments. In K. Tobin (Ed.), *The practice of constructivism in science education* (pp. 171-180). Hillsdale, NJ: Lawrence Erlbaum Associates.

Geelan, D.R. (1997). Epistemological anarchy and the many forms of constructivism. *Science and Education*, 6: 15-28.

Good, R. (1993). The many forms of constructivism. *Journal of Research in Science Teaching*, 30(9): 1015.

Howard, R.W. (1987). *Concepts and schemata: An introduction*. Philadelphia, PA: Cassell Education.

Kelly, G.A. (1955). *The psychology of personal constructs*. New York: Norton Press.

Marton, F., Dall'Alba, G., and Beaty, E. (1993). Conceptions of learning. *International Journal of Educational Research*, 19: 277-285.

Marton, F. and Saljo, R. (1976). On qualitative differences in learning. I: Outcome and process. *British Journal of Educational Psychology*, 46: 4-11.

Matthews, M.R. (1993). Constructivism and science education: Some epistemological problems. *Journal of Science Education and Technology*, 2(1): 359-370.

Noddings, N. (1990). Constructivism in mathematics education (Monograph 4). *Journal for Research in Mathematics Education*, 7-18.

Osborne, R.J. and Wittrock, M.C. (1983). Learning science: A generative process. *Science Education*, 67(4): 489-508.

Osborne, R.J. and Wittrock, M. (1985). The generative learning model and its implication for science education. *Studies in Science Education*, 12: 59-87.

Pfundt, H. and Duit, R. (1994). *Bibliography: Students' alternative frameworks and science education* (Fourth edition). Kiel, Germany: Institute for Science Education.

Piaget, J. (1970). *Science of education and psychology of the child*. New York: Grossman.

Piaget, J. (1972). *The principles of genetic epistemology*. New York: Basic Books.

Pines, L.A. and West, L.H.T. (1983). A framework for conceptual change with special reference to misconceptions. In J.D. Novak (Ed.), *Proceedings of the International Seminar in Science and Mathematics* (pp. 65-71). Ithaca, NY: Cornell University Press.

Solomon, J. (1987). Social influences on the construction of pupils' understanding of science. *Studies in Science Education*, 14: 63-82.

Sutherland, P. (1992). *Cognitive development today: Piaget and his critics*. London, UK: Paul Chapman Publishing.

Tobin, K. (1990). Social constructivist perspectives on the reform of science education. *Australian Science Teachers' Journal*, 36(4): 29-35.

Toohey, S. 1999. *Designing courses for higher education*. Buckingham, UK: Open University Press.

Watts, M. (1994). Constructivism, re-constructivism and task-orientated problem-solving. In P.J. Fensham, R.F. Gunstone, and R.T. White (Eds.), *The content of science: A constructive approach to its teaching and learning* (pp. 39-58). London, UK: The Falmer Press.

Chapter 9

Application of ICT to Provide Feedback to Support Learning in First-Year Science

Mary Peat
Sue Franklin
Charlotte Taylor

INTRODUCTION

University teaching appears to have been reinvented during the past decade in response to comment and criticism from the major stakeholders in the process. Worldwide university administration, in response to governmental dicta, has also argued for implementation of transparent processes, accountability of degree programs, and for higher retention and progression rates of students. For example, in Australia, national CEQ (course experience questionnaire) data are now driving quality assurance processes, and in turn universities are looking closely at the inputs and products of teaching and learning. In their benchmark Australian study on the first-year experience, Mc-Innis, James, and McNaught (1995) identified some serious deficiencies in the student experience and suggested that these may be related to the size of Australian first-year classes.

In a follow-up study, McInnis, James, and Hartley (2000) indicated that although Australian universities have attempted to address many of these deficiencies, the student experience is still not always a positive one. In particular, both the 1995 and 2000 studies point toward the provision of relevant and timely feedback as being an important issue for students. The McInnis, James, and Hartley (2000) study highlights that, in comparison with the previous study, there is a general student perception that fewer staff are available to discuss their work, and this is reflected in their perceptions of the extent to which staff usually give feedback on student progress. This chapter looks at the use of information and communications technologies (ICT) in providing relevant and timely feedback to help support the learning activities of a large first-year student group at the University of Sydney.

DIVERSITY OF AUSTRALIAN
FIRST-YEAR SCIENCE STUDENTS

The teaching of science to very large numbers of first-year students is a characteristic of the on-campus learning environment at the larger Australian universities (see Chapter 10). For example, at the University of Sydney, science subjects are taught to up to 2,000-first year students each year, from faculties as diverse as agriculture, arts, economics, education, engineering, nursing, pharmacy, and science, with incoming students having a wide range of academic abilities, differing backgrounds in the subject area, and diverse incoming generic skills and motivation. Many of the students are in "service" subjects progressing into professional higher-year courses, which may not appear to students to be closely related to mainstream science. Many of the students are enrolled in professional-degree programs and as such are highly motivated and know where they are going, but a large proportion are enrolled in the generalist science degree, which allows them a wide choice of subjects but often means they are unsure of their future directions, which is evidenced by fluctuating levels of motivation and interest in the subjects (see Chapter 7).

Another important issue, as McInnis, James, and McNaught (1995) report, is that some students travel long distances, and many live on relatively low incomes, often paying significant proportions of their incomes in rent, requiring them to be in paid employment. It is apparent that, in the current economic climate, many students have to juggle university commitments with employment, potentially missing some of the structured teaching and learning sessions and, more important, not being able to take advantage of campus-based course materials and face-to-face assistance from staff. Demographic data from first-year biology students at the University of Sydney show that students not only are taking a full-time university load but are working long hours in paid work (Peat and Franklin, 2002), potentially making it extremely difficult for them to fulfill course expectations.

SUPPORTING FIRST-YEAR SCIENCE STUDENTS WITH RELEVANT FEEDBACK

Recent longitudinal studies of first-year students at Australian campuses indicate that our students may need more support than their predecessors due in part to the increasing heterogeneity of the student cohort (McInnis, James, and McNaught, 1995; McInnis, James, and Hartley, 2000). These large first-year science courses generally involve repeat lecture series, multiple concurrent laboratory sessions, seemingly never-ending reports to mark, and vast numbers of examination papers to grade, and academics are facing many difficulties in their endeavors to support student learning. Some of these difficulties arise from the increasing number and previously mentioned diversity of students, along with the reduction in recurrent resources. In particular there has been a reduction in tenured staff numbers, with a concurrent increase in the employment of casual staff for teaching. This increased casualization of teaching staff in science practicals and tutorials has led to issues related to commitment of resources for staff training and the quality of the teaching. In addition, with more casual teachers involved comes a reduction in the interactions between permanent academic staff and students, all of which results in less support available for students outside of formal class time. In particular, this reduction in support may potentially result in a lack of feedback to students on their performance during the crucial early stages of a university program.

It is possible for students to receive feedback on their progress via many avenues, both online and offline, and including formative and summative assessment tasks, during discussions and during peer group activities. More recently, feedback issues have been addressed by the development of online assessment resources, particularly when dealing with large student numbers. Such resources are available for use by students anytime and anyplace. In this way, large groups of students can be provided with the feedback they are now requesting, and this may help them in their final assessment tasks. The development of online assessment resources has a number of advantages over offline pen-and-paper tasks. They can be easily marked, provide instant feedback, and can be taken repeatedly by students in order to assess and improve performance. Also, online tests can be taken unsupervised in students' own time (for further discussion of these advantages see Bugbee and Bernt, 1990, and Bugbee, 1996). Clariana (1993) has shown that online assessments allow students to tailor their use to their own learning style, while Zakrzewski and Bull (1999) emphasize the advantages of online assessment in terms of fast feedback to large numbers of students with no staff involvement. They also indicate that the use of online formative assessment prior to summative tests reduces student anxiety. The contribution of formative computer-based assessment on improvements in student learning outcomes is documented by Buchanan (2000), who found that undergraduate psychology students who used an online formative assessment package, which provided instant feedback, performed better in the end-of-course summative assessment than those who did not use the package. Gretes and Green (2000) also found a positive relationship between the number of practice tests taken and the final course grade when they supplemented a lecture course with online formative quizzes. These results suggest that engaging Web-based interactions facilitate learning and that such interactions are key components in an e-learning environment as advocated by Bork (2001).

The educational research literature also shows that students who make use of every learning opportunity approach the final assessment tasks with a greater likelihood of high performance outcomes (e.g., De Vita, 2001; Heffler, 2001). Fowell, Southgate, and Bligh (1999) suggest that student learning is best served by the provision of a diverse range of assessment (and thus feedback) methods, as individual methods, exclusively employed, may disadvantage some students. They also suggest that, from the teacher perspective, using a range of methods allows performance from different sources to be related. Seale, Chapman, and Davey (2000), who investigated which types of assessment students found most motivating for their learning, found that having a range of opportunities for receiving feedback on performance was most motivating. They also found that the relevance and content of the assessment appeared to influence student motivation, as well as the enthusiasm of the teachers. Feedback on performance, especially that of a formative nature, has been shown to be a valuable tool in the learning process, enabling students to assess their own progress and understanding and remedy any weakness exposed by the assessment (Peat and Franklin, 2003; Clariana, 1993; Macdonald, Mason, and Heap, 1999; Zakrzewski and Bull, 1999). However, for this feedback to be effective it needs to be provided early in the learning process (Brown and Knight, 1994; McInnis, James, and McNaught, 1995) and have some degree of prescription about how to improve performance (Wiliam and Black, 1996). Fowell, Southgate, and Bligh (1999) argue that the presentation of summative grades also requires the provision of effective feedback to students, in both summative and formative assessment tasks, enabling students to identify their strengths and weaknesses so they can improve future performance. The more recent study of McInnis, James, and Hartley (2000) indicates that lack of suitable, timely, and relevant feedback is still a feature of many university courses in Australia and emphasizes that Australian students have higher expectations of support than their predecessors. This may be a reflection of the increasing diversity and numbers of students attending tertiary institutions. As well as addressing this issue, the move toward provision of online feedback will encourage students to be self-motivated.

Within the scientific academic community, many of us have been striving to provide alternative learning experiences to support students in the face of increasing student numbers and reduced staffing resources. Over the past decade many of us have increasingly used IT to help bridge the widening gap between acceptable and affordable provision of support. As teachers, many of us have introduced new and varied offline and online materials with their inherent flexibility of anywhere/anytime access to support student learning and enhance the learning experience.

One urgent issue for science teaching is the need for students to receive timely feedback during the writing process and in the context of undergraduate assignments and learning. This has been acknowledged both at the policy level of the university and the learning level of individual courses (Lea and Street, 1998). In an attempt to improve the environment of giving and using feedback, curriculum design has given it a more explicit and integrated focus in the development of writing skills. A traditional linear approach to writing has thus given way to a more cyclical approach, in which the learning of scientific writing uses feedback at all stages of the process (Taylor and Drury, 1996). However, students are usually working alone during the key period of preparation of drafts and final reports, and at this point they frequently feel that they need individual feedback on their efforts. This feedback is also often required at short notice, so that providing appropriate help is a particular challenge for staff working with a large student cohort. Using the Internet is one way to provide this support, especially as the answers to questions from one student can be shared by many students accessing the same resource. With respect to the use of IT-based feedback on assignments, Breen and colleagues (2001) have reported that IT-based learning modes may be perceived by students as decreasing the interaction with teachers and providing correspondingly fewer opportunities for feedback on assignments. They have also documented students' facility and familiarity with using IT in their learning. The combination of discussing and learning writing online has been documented in the context of small courses and those teaching postgraduates, and as assessable parts of the curriculum Lea (2001). This study also evaluated the inherent learning opportunities offered by this new medium as the field of writing development expands to incorporate mixtures of informal writing and discussion via e-mail and the more formal requirements of written assignments. Coupled with this is the potential to use an online forum for giving feedback on writing, albeit as part of a distance-education program (Weedon, 2000).

PROVISION OF ONLINE FEEDBACK TO FIRST-YEAR BIOLOGY STUDENTS AT THE UNIVERSITY OF SYDNEY

For almost a decade at the University of Sydney we have been providing large groups of first-year, undergraduate biology students with computer-based resources to support them in their learning. Since 2000, the materials have been presented via a virtual learning environment (Figure 9.1) that allows students easy access to all available learning resources <http://FYBio.bio.usyd.edu.au/VLE/L1/>. This is also described elsewhere (Peat, 2000).

In particular, a major challenge at the University of Sydney has been to improve the feedback given to students in a large first-year science class (1,700 biology students in 2003) so that students had ongoing opportunities to self-appraise their performance during the course and reflect on their learning. The increased size of the student

FIGURE 9.1. Entry into the Virtual Learning Environment

body has motivated us to develop teaching and learning opportunities to support student diversity and heterogeneity with respect to prior knowledge of the discipline, personal professional goals, and personal motivation for learning. Further motivation came from the knowledge that many of our students, while enrolled in a full-time degree program, are working up to twenty hours on a casual basis every week (Peat and Franklin, 2002), a figure supported nationally and across all undergraduate years by the work of McInnis, James, and Hartley (2000). Our goals were to provide the best supportive learning environment that was possible.

In biology at the University of Sydney, we have created a culture of giving ongoing support and feedback on the learning process using multiple media. Feedback is provided during weekly laboratory classes (e.g., online summative quizzes, crossword puzzles, self-test quizzes in the laboratory notes) and in learning modules and self-assessment modules via our virtual learning environment. Our research on the perceptions of the usefulness of our resources in giving feedback indicates that each resource is valued for different reasons (Peat and Franklin, 2002, 2003; Franklin, Peat, and Lewis, 2003). In particular, students indicate that the formative assessment resources provide them with feedback that aids in revision, learning new knowledge, and consolidating knowledge (Peat and Franklin, 2003).

In reference to the provision of feedback to students, we consider our flagship activities to be the use of feedback, both online and offline, to develop scientific writing skills, and the provision of computer-based self-assessment modules for personal appraisal of performance.

USE OF FEEDBACK IN THE DEVELOPMENT OF SCIENTIFIC WRITING SKILLS

A writing learning cycle was developed for use with first-year students taking the first-semester biology course and implemented within the laboratory component of the course so that all discussions, tasks, and assessments were facilitated by the laboratory demonstrating staff (Taylor and Drury, 1996). The writing learning cycle thus had to be communicated and modeled first for the staff during hands-on training sessions before they progressively introduced it to students during the semester.

The Writing Cycle

The writing cycle was integrated with discipline-based tasks in the laboratory classes to maximize efficient use of time and to emphasize the links between communicating through writing and understanding the biological concepts, as indicated by Lea and Street (1998). The writing cycle (Figure 9.2) consisted of five stages: planning and preparation, writing, feedback, reflection, and moving to the next task.

Working in groups, our students were first given the task to review two pieces of writing, to develop a set of criteria for assessing good writing, and then to use these criteria to appraise an individual piece of writing from a colleague. This piece of writing comprised the results and discussion sections of a laboratory report on a class investigation on the number of slaters found in two different types of leaf litter. Having completed the written critiques the student group was asked to reflect on the reports and discuss among themselves the similarities and differences in the various approaches to writing. In addition, the teacher gave written feedback on these practice reports during the same class.

The next class investigation (an experiment to determine the quality of drinking water) was also written up individually by the students

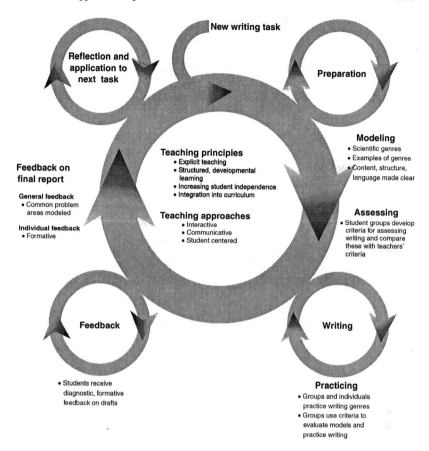

FIGURE 9.2. The Learning Cycle

and a draft handed in during a laboratory class, at which time they received written formative feedback from their teacher. The final report was handed in two weeks later for final summative assessment as well as written formative feedback.

Supporting the Writing Cycle with an Online Discussion Seminar

During a three-week period, while this draft report writing, feedback process, and final writing was going on, the course coordinator moderated an online seminar on the process of writing the report. The Webteach program (Hughes and Hewson, 1998) was used, as it allowed an appropriate degree of teacher moderation for such a discussion and the asynchronous mode encouraged questions and discussions.

The use of the online seminar has been enthusiastically embraced by students, the majority accessing it at some time while writing their report, as shown in Table 9.1. Activity is measured by movements around the site by individuals.

In 2001, approximately 7 percent of the students accessing the site participated in the discussions, a figure which compares very favorably with other studies (Lea, 2001), particularly since participation was not obligatory. These students asked questions of the moderator and other students in a number of threaded discussions based loosely on the different areas of the report, e.g., results, discussion, references, appendix, and statistical test. Although the moderator provided definitive answers to most questions, other questions were developed by the students as they compared experiences and results and worked together to solve problems.

Research Findings and Learning Outcomes

The online seminar has been evaluated in 2000 and 2001 using paper-based qualitative and quantitative surveys of 200 students, as well as reflective discussions from the moderator. Usage statistics from the program also provided data on access rates, times, and duration for individuals and the cohort as a whole. The usage data identified students as participants or "lurkers" (Pearson, 2000) and allowed a detailed analysis of patterns of activity and learning for individual participants. The majority of students used the online seminar at home and were generally happy with access and with navigating the

TABLE 9.1. Student Use of the Online Seminar During a Two-Year Period

Year	n	Total visitors	Visitors who participated	Activity
2000	1,100	283	28	18,000
2001	1,100	784	73	40,000

site. Ninety-five percent of those visiting the seminar used information from the discussions in their reports. Distinct peaks of use were found during each twenty-four-hour period, including a series of smaller ones associated with accessing during the day while on campus, with an increase in use as the evening progressed. Students commonly logged on late in the evening, presumably while they were working on their report, and a subset remained active between midnight and 3 a.m.

A range of reasons was given for visiting and participating in the seminar. These included the following:

"Getting personal attention for my question"
"Helped me get started"
"Can ask questions and get answers anytime"
"Read the answers while I'm writing my report"
"Realize that others have the same problems"
"Helps me to understand through discussing"

The design of the discussion mode (Hughes and Hewson, 1998) makes students read through and engage with each discussion as it develops rather than skipping between the most recent responses. The system also allows those students less confident in articulating their problems to lurk and gain information or understanding (Pearson, 2000). Overall, the seminar allows the moderator to provide feedback to individuals while being accessible to all students, thus providing more equitable access to all students. The asynchronous nature of the interaction has also encouraged a new level of independence in problem solving and reflecting on problems and solutions, as suggested by Dysthe (2002).

Feedback is an inherent component of the writing process and as such is integral to the learning cycle adopted by the biology writing program. However, a perception common to both students and staff has been that feedback is part of a linear progression following the completion of assignments. The writing cycle described here aims to encourage reflection on feedback during discussions, practice assignments, and draft reports. A feedback session for a draft report has become part of the cycle and has proved very successful in giving each student individual attention. It has changed the focus of the task, whereby students no longer leave their writing until the last moment, since rewards are offered for the submission of drafts, and they have an opportunity to reflect on their work before handing in a final report. The extent to which students change and develop their work following feedback is not yet clear.

USE OF COMPUTER-BASED SELF-ASSESSMENT MODULES PROVIDING FEEDBACK

Of all our computer-based materials that give feedback to first-year biology students, we consider the self-assessment modules (SAMs) to be the most innovative of our teaching and learning materials. The SAMs are designed to draw together related parts of a course to help students make connections between topics in biology and to promote a deeper learning strategy, while providing an enjoyable feedback and reinforcement session. They are additional, optional materials designed to let students identify their own levels of understanding. The SAMs are located in our virtual learning environment <http:// FYBio.bio.usyd.edu.au/VLE/L1/ResourceCentre>, where currently twenty-one SAMs are available to first-year biology students.

Educational Rationale for Provision of SAMs

While the courses are thematic, the SAMs are organized around specific content; thus students are taken down a lateral pathway to encourage them to see the relationship between the materials. The first SAM was designed and developed, using tailor-made templates, and introduced in 1997. Subsequent to this, several SAMs have been produced each year, with content being entered into these question templates. Each SAM tests the student on four levels of increasing difficulty, using Bloom's (1956) *Taxonomy of Educational Objectives* as a guide to develop the levels. Thus the content of the questions can be reused (from level to level) but with an increasing cognitive requirement, and appropriate question types have been developed for each level of difficulty. Level 1 tests *content and knowledge* with the use of multiple-choice questions and drag-and-drop scenarios, but with the answer always on the screen. Level 2 tests *application* of content using some multiple choice, but mostly with a format that expects text input from the student. Level 3 tests *analysis* and uses question formats, as with Level 2, but with the addition of two-part questions and formats requiring the building up of diagrams, flowcharts, and so on. Level 4 tests *synthesis* of information, the most-used format being free-flow prose, where the student is expected to synthesize information in response to a question. This format is not computer marked but self-assessed by the students, who compare their work with sample answers, with the option of self-scoring their own performance. A more detailed description of the educational design of the SAMs is contained in Peat (2000). See Figures 9.3, 9.4, and 9.5 for specific illustrations of these questions.

At the beginning of each SAM, students are directed to a statement of educational rationale, in particular about the value of self-assessment/appraisal in achieving learning outcomes. Students are informed that each module is presented on four levels of difficulty and what each level is testing. When they quit a SAM, students are asked to review their performance and consider what their results indicate to them in terms of their learning. To do this they compare their performance in the SAM with that of a fictitious student. They are given the

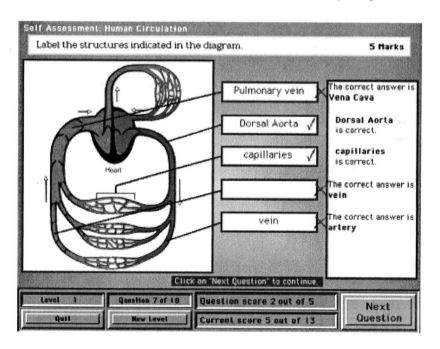

FIGURE 9.3. Level 1 SAM Question

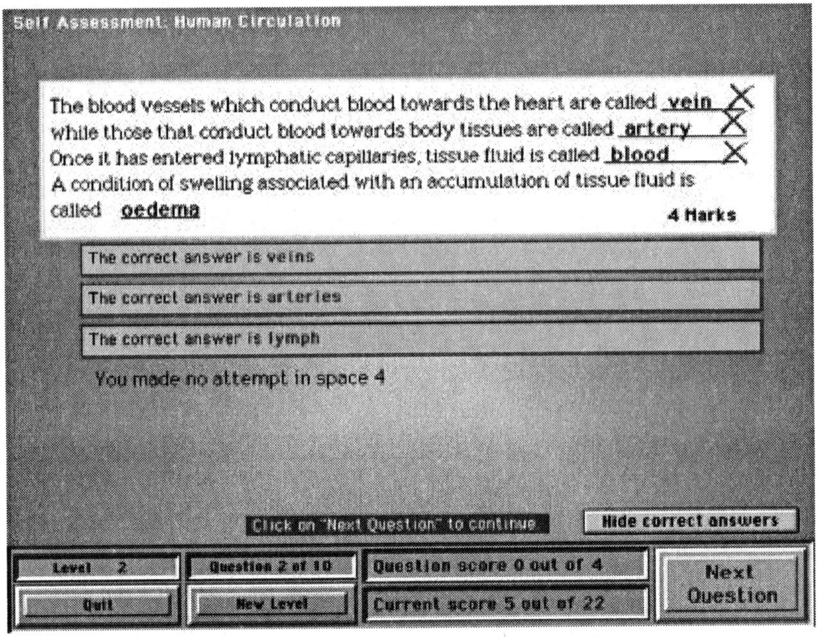

The blood vessels which conduct blood towards the heart are called **vein** X
while those that conduct blood towards body tissues are called **artery** X
Once it has entered lymphatic capillaries, tissue fluid is called **blood** X
A condition of swelling associated with an accumulation of tissue fluid is
called **oedema**

4 Marks

The correct answer is veins

The correct answer is arteries

The correct answer is lymph

You made no attempt in space 4

Click on "Next Question" to continue | Hide correct answers

| Level 2 | Question 2 of 10 | Question score 0 out of 4 | Next Question |
| Quit | New Level | Current score 5 out of 22 | |

FIGURE 9.4. Level 3 SAM Question

example of a fictitious student, Mary Rotelearner, whose performance is satisfactory at Levels 1 and 2 but is very poor at Levels 3 and 4. It is explained that, with results like this, Mary Rotelearner has been able to deal with content and knowledge (rote learning) but is not very good at analysis and synthesis (deep learning) of the materials. In the log-out information, students are encouraged to reflect on their performance at each level and to ask themselves what type of learning strategy they are adopting and whether it is appropriate.

Research Findings and Learning Outcomes

The SAMs have been evaluated on an ongoing basis since their introduction in 1997, using both paper-based and online surveys, and utilizing both qualitative and quantitative methodology and focus-group discussions. Originally the developers did not intend to collect usage statistics from students, only formative evaluation information

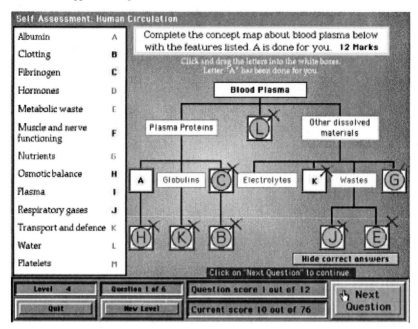

FIGURE 9.5. Level 4 SAM Question

that would potentially help improve the product. The emphasis was on student perceptions of the usability and usefulness to learning of the SAMs.

A paper-based survey in 1997, investigating how students were using the first SAM (i.e., the prototype), showed that most students using this resource did so alone and were not very likely to complete it (see Table 9.2). As more SAMs were introduced it was decided to evaluate them electronically as the students logged out from each individual module. Comparison across the years, using these online surveys and asking the same questions for all SAMs, shows a correlation with the 1997 preliminary data (Table 9.2). It is clear that the majority of students using the SAMs do so on their own, probably at home, and do not complete the entire module in one sitting. They also value being able to choose a level of difficulty from within the SAM when self-assessing. This is consistent with the design of this type of learning resource, in that it offers flexibility both for access and order

TABLE 9.2. Student Feedback About the SAMs, 1997-2001, in Percentages

Response	1997	1998	1999	2000	2001
Used alone	81	96	99	96	100
Completed a SAM	50	26	22	45	65
Enjoyed using the module	100	87	87	100	94
Like to choose level to do	–	96	94	96	94

of use, and that a student can quit the module at any time during its use but still get formative feedback on performance. The data for 1998 and 1999 show the students did not enjoy the modules as much as in the other years, and fewer of them completed the modules. During this time there were ongoing technological changes to delivery systems as more first-year biology materials were being made available via the Internet, which resulted in some teething problems associated with the downloading of some of the modules. The provision of the First Year Biology CD-ROM from 2002 onward, containing all learning modules and self-assessment modules, has removed this difficulty.

Students were asked open-ended questions about how the SAMs helped them in their understanding of content and their learning generally. When categorized, the responses indicated that feedback helped in revising, in understanding the material, and in indicating the areas that need improvement (Peat and Franklin, 2002). The feedback that offered useful diagrams and diagrammatic representation of ideas and adapted different approaches from the textbook was rated highly by students.

In a structured set of survey questions using a Likert five-point survey, with 5 representing strongly agree and 1 representing strongly disagree, high mean scores were obtained in the areas of assessing understanding, testing concepts, relating concepts, and giving useful feedback (Table 9.3). Overall, the modules have maintained a highly favorable response from the students through the years.

CONCLUSION

We believe that teaching and learning strategies which encompass providing timely and relevant feedback and include both online and

TABLE 9.3. Perceptions of the Usefulness of SAMs to Student Learning As Indicated by Likert Means

Response choices	1998	1999	2000	2001
Help assess understanding	4.30	4.25	4.39	4.53
Help test concepts	3.91	4.21	4.30	4.24
Help relate concepts	3.57	3.66	4.17	3.50
Give useful feedback	4.09	4.26	4.52	3.83
Overall rating	4.09	4.15	4.48	3.94

offline and formative and summative activities are essential for good learning outcomes. These activities may be supervised in classroom situations or left as independent student tasks. In biology at the University of Sydney, the feedback is provided in a range of activities including quizzes, mock exams, written reports or essays, group posters, oral presentations, remedial programs, and self-assessment tests. The important issue for us is that there be a mix of opportunities for students to gain appropriate feedback embedded in our courses. We consider that in all first-year courses it is important that students are supported in their transition from secondary-school-based learning, to university-based learning and this may need a special emphasis on early and relevant feedback, especially during the first semester of their university experience.

There has been a strong trend toward giving better feedback to students, especially in their early months at university. This has been encouraged in Australia by science-specific activities (such as the annual conference of UniServe Science <http://science.uniserve.edu.au>, the First Year Experience conferences hosted by Queensland University of Technology <http://www.qut.edu.au/talss/fye/home.htm>, and by the various conferences, both national and international, on teaching and learning issues). In addition, the government initiatives such as those of the Australian Universities Teaching Committee (AUTC), have fostered an awareness in the academic community toward the improvement of the learning experiences of students. We still have some distance to travel, but we are on the right road.

REFERENCES

Bloom, B.S. (1956). *Taxonomy of educational objectives:* Handbook 1, *Cognitive domain.* New York: Longman.

Bork, A. (2001). What is needed for effective learning on the Internet? *Educational Technology and Society,* 4: 3. Available online at <http://ifets.ieee.org/periodical/vol_3_2001/bork.html>.

Breen, R., Lindsay, R., Jenkins, A., and Smith, P. (2001). The role of information and communication technologies in a university learning environment. *Studies in Higher Education,* 26(1): 96-114.

Brown, S. and Knight, P. (1994). *Assessing learners in higher education.* London: Kogan Page.

Buchanan, T. (2000). The efficacy of a World-Wide Web mediated formative assessment. *Journal of Computer-Assisted Learning,* 16: 193-200.

Bugbee, A.C. (1996). The equivalence of paper-and-pencil and computer-based testing. *Journal of Research on Computing in Education,* 28: 282-299.

Bugbee, A.C. and Bernt, F.M. (1990). Testing by computer: Findings in six years of use 1982-1988. *Journal of Research on Computing in Education,* 23: 87-100.

Clariana, R.B. (1993). A review of multiple-try feedback in traditional and computer based instruction. *Journal of Computer Based Instruction,* 20(3): 67-74.

De Vita, G. (2001). Learning styles, culture, and inclusive instruction in the multicultural classroom: A business and management perspective. *Innovation in Education and Training International,* 38(2): 165-174.

Dysthe, O. (2002). The learning potential of a Web-mediated discussion in a university course. *Studies in Higher Education,* 27(3): 339-353.

Fowell, S.L., Southgate, L.J., and Bligh, J.G. (1999). Evaluating assessment: The missing link? *Medical Education,* 33(4): 276-281.

Franklin, S., Peat, M., and Lewis, A. (2003). Non-traditional interventions to stimulate discussion: The use of games and puzzles. *Journal of Biological Education,* 37(2): 79-84.

Gretes, J.A. and Green, M. (2000). Improving undergraduate learning with computer-assisted assessment. *Journal of Research on Computing in Education,* 33(1): 46-54.

Heffler, B. (2001). Individual learning style and the learning style inventory. *Educational Studies,* 27(3): 307-316.

Hughes, C. and Hewson, L. (1998). Online interactions: Developing a neglected aspect of the virtual classroom. *Educational Technology* 38(4): 48-55.

Lea, M.R. (2001). Computer conferencing and assessment: New ways of writing in higher education. *Studies in Higher Education,* 26(2): 163-181.

Lea, M.R. and Street, B.V. (1998). Student writing in higher education: An academic literacies approach. *Studies in Higher Education,* 23(2): 157-172.

Macdonald, J., Mason, R., and Heap, N. (1999). Refining assessment for resource-based learning. *Assessment and Evaluation in Higher Education,* 24(3): 345-354.

McInnis, C., James, R., and Hartley, R. (2000). *Trends in the first-year experience in Australian universities.* [A project funded by the Evaluations and Investigations Programme, Higher Education Division, DETYA.] Canberra: Australian Government Printing Service. Available online at <http://www.detya.gov.au/archive/highered/eippubs/eip00_6/fye.pdf>.

McInnis, C., James, R., and McNaught, C. (1995). *First year on campus: Diversity in the initial experiences of Australian undergraduates.* [A commissioned project for the Committee for the Advancement of University Teaching.] Canberra: Australian Government Printing Service.

Pearson, J. (2000). Lurking, anonymity and participation in computer conferencing. In *Communications and networking in education, international federation for information processing* (1999, Aulanko, Finland).

Peat, M. (2000). Online self-assessment materials: Do they make a difference to student learning? *Association for Learning Technology Journal,* 8(2): 51-57.

Peat, M. and Franklin, S. (2002). 2002 survey of demographics of first-year biology students, including casual hours worked in semester. Available online at <http://fybio.bio.usyd.edu.au/fyb/tdg/AnnReports/wk_sur02.htm>.

Peat, M. and Franklin, S. (2003). Has student learning been improved by the use of online and offline formative assessment opportunities? *Australian Journal of Educational Technology,* 19(1): 87-99.

Seale, J., Chapman, J., and Davey, C. (2000). The influence of assessments on students' motivation to learn in a therapy degree course. *Medical Education,* 34(8): 614-621.

Taylor, C.E. and Drury, H. (1996). Teaching writing skills in the science curriculum. *Research and Development in Higher Education,* 19: 160-164.

Weedon, E. (2000). Do you read this the way I read this? *British Journal of Educational Technology,* 31(3): 185-197.

Wiliam, D. and Black, P. (1996). Meanings and consequences: A basis for distinguishing formative and summative functions of assessment? *British Educational Research Journal,* 22: 537-538.

Zakrzewski, S. and Bull, J. (1999). The mass implementation and evaluation of computer-based assessments. *Assessment and Evaluation in Higher Education,* 23(2): 141-152.

Chapter 10

Assessing for Learning in the Crucial First Year of University Study in the Sciences

Frances Quinn

INTRODUCTION

Student diversity is becoming increasingly characteristic of first-year classes and is one of the imperatives for more innovative assessment in tertiary education (James, McInnis, and Devlin, 2002). Among first-year students there is variation in age, enrollment type (on or off campus), and language background, and growing numbers of first-year students are juggling part-time paid employment with full-time study (McInnes, James, and Hartley, 2000). Numerous first-year students at some Australian universities are from socioeconomically disadvantaged environments, and may be hampered by previous systemic educational disadvantage; dependent, passive learning habits; and the lack of "cultural capital" described by Muldoon (see Chapter 7). Many science students in this era of mass tertiary education do not have intrinsic interest in science "for its own sake" (Laws, 1996, p. 25). This has clear implications for student motivation, especially in first-year science classes. These are often "service classes," that is, large, compulsory, introductory classes prerequisite to later more specialized areas of study, and therefore potentially of low perceived relevance to students' aspirations or interests. Increasingly, in Australia (Niland, 1998) and the United Kingdom (Dunbar, 1995), students are entering science programs having taken less-intensive, generalist science options in secondary school. These aspects of student diversity are central to the issue of assessment in first year, as relative youth, time pressures, lack of intrinsic interest, and limited background knowledge are frequently associated with ineffective reproductive orientations to learning, which may either be entrenched or challenged by assessment choices in first year.

Awareness of the crucial importance of the first-year experience is growing in Australia and beyond (McInnes, James, and Hartley, 2000; Pitkethly and Prosser, 2001). In their transitional first year, students form views on what university learning is about that will influence their learning behaviors and expectations in subsequent years. First year is also a period of high risk for student failure and attrition. Harnessing the potential of appropriate assessment strategies to contribute to student learning and retention in first year is therefore crucial, as assessment is "the most powerful lever teachers have to influence the way students respond to courses and behave as learners"(Gibbs, 1999 p. 41). Yet despite the need for assessment that is centered on the requirements of this diverse group of learners, first-year science classes traditionally have been characterized by assessment techniques least supportive of student learning and motivation, such as heavy reliance on end-of-semester objective exams. In part this is because of the time and financial constraints associated with large class sizes. Indeed, assessment issues are considered one of the most problematic aspects of large classes (Gibbs, 1999).

The aim of this chapter is to investigate the place of assessment within the current pedagogical context of first-year science. Some general dimensions of assessment in tertiary science education are described, followed by discussion of the relationship between assessment and first-year science curriculum issues and the influence of assessment practices on the approaches students take to their learning. Some general principles for assessment of first-year students are then presented, along with examples drawn from available published case studies of assessment alternatives used in a variety of large introductory science classes. Ultimately, it is hoped that this chapter demonstrates the importance of assessment innovation for first-year science students and shows that this can be achieved within resource constraints, while providing a source of ideas and further information for first-year tertiary science teachers.

DIMENSIONS OF ASSESSMENT

Assessment does "several things at once" (Ramsden, 1992) in several different ways, and some of these dimensions of assessment are summarized in Table 10.1. Few of these dimensions of assessment are mutually exclusive, for example, there are no formal assessments that are solely summative or formative, and these categories occupy opposite ends of a continuum (Brown, 1999).

Tensions about purpose, validity, and reliability pervade choices of assessment strategy. For instance, reliable assessment tasks such as multiple-choice examinations have a predominantly summative role, with little formative potential and low validity. That is, they provide little feedback to students and limited information about the gamut of student learning relevant to a particular real-world context. There is a widespread movement in the tertiary context to view assessment from

TABLE 10.1. Different Types of Assessment

Assessment type	Definition	Example
Summative	The purpose is to measure learning and grade students.	Examination
Formative	The purpose is to contribute to student learning by providing feedback on meaningful learning tasks (Yorke, 2001).	Presubmission assignment critique
Formal	Tasks are submitted primarily for summative or formative assessment of student progress.	Laboratory report
Informal	Assessment and feedback on student progress occurs within routine learning/ teaching interactions.	Demonstrator commenting on titrating technique
Criterion referenced	This type measures learning and assigns grades according to the extent to which a student meets predefined criteria.	Criterion-based assignment feedback sheet
Norm referenced	This type measures learning and assigns grades according to the performance of a student relative to other students of the same or similar cohorts.	Grading to a bell curve
Authentic	Assessment tasks relate to "real-world" requirements. Often these are complex and cross a variety of learning objectives.	Assessment component of problem-based learning

a more learner-centered perspective, by emphasizing more the formative goals of assessment (e.g., Gibbs, 1999; Yorke, 2001). Formative assessment emphasizes timely, relatively frequent, high-quality feedback that is essential for learning but can be expensive to deliver, particularly in large first-year science classes (over 150 students) sustaining costly laboratory sessions (Peat, 2000). Assessment validity is also receiving more emphasis, in tandem with criterion-referenced assessment (Brown and Knight, 1994) and more authentic assessment strategies.

The Impact of Assessment on What Students Learn

Choosing which mix of assessment types to use depends on what learning is to be assessed, and this is inextricably linked to the perceived role of first-year introductory classes. One of the most powerful messages from students is that formal, summative assessment defines the curriculum for them by structuring their allocation of time and effort (Rowntree, 1987), particularly when they are operating under time constraints (Gibbs, 1999). Therefore, if first-year science curricula incorporate learning goals such as transferable generic skills (including skills of self-evaluation), higher-order thinking skills, and an understanding of how to "do" science, assessment needs to be aligned with these goals for students to take them seriously.

The need for assessment to be aligned with learning goals highlights the difficult question of establishing the appropriate balance between content, scientific and generic skills, and attitudinal factors in introductory science curricula. Concerns have been expressed for some years about a detrimental focus on content over scientific and generic skills in tertiary science teaching, particularly in introductory courses (e.g., Hegarty-Hazel, 1990). One prevalent perception of the role of first year in science is apparent in the following quotation: "In the sciences you need an information pack before you can start to discuss anything, and those tools may have to be acquired by rote learning before we can do anything with them" (Dunbar, 1995, p. 181).

Similar views have been expressed elsewhere (e.g., Entwistle and Marton, 1984), and the hierarchical structure of scientific knowledge may indeed require initial attention to basic concepts and procedures. However, a problem frequently associated with this "building block" focus on equipping students with a broad factual knowledge base is transmission of excessive content in lectures and recipe-following illustrative laboratory sessions. This type of first-year science environment suffers from the "custom-built irrelevance" criticized by Lowe (1994) for stifling student interest and hampering learning. It is increasingly being argued that in introductory classes, as well as those in later years, it may be necessary to "unstuff" the curriculum of excessive content and develop more student-centered pedagogical techniques to foster broader scientific and generic skills. Such changes to first-year science curricula are occurring internationally, despite some resistance to reduction of content (e.g., Passey, 1996; Stokstad, 2001).

In addition to scientific and generic skills, the goals of first-year science education should also include attitudinal factors, such as student interest and scientific curiosity, and an ethical approach to scientific research. If attitudinal factors are accepted as part of the curriculum, they also must be assessed in some way: "we cannot teach without assessing attitudes" (Rowntree, 1981, p. 190).

These shifts of curriculum focus and associated pedagogical innovations are bringing with them requirements for changing assessment practices. Unless assessment is aligned with curriculum goals and integrated with pedagogical approaches, the potential benefits of these teaching innovations may not be realized. Recognizing the potential power of assessment in driving student learning has far-reaching effects on curriculum design (Yorke, 1998), making assessment tasks central rather than accessory to other curriculum decisions.

The Role of Assessment in How Students Learn

Assessment is crucial in the first year because it defines *what* students learn, but it also directly shapes *how* students learn by influencing their approaches to learning. Compelling evidence indicates that students adopt either a deep or surface learning approach, depending on their perceptions of the requirements of a particular learning context (for reviews see Prosser and Trigwell, 1999, or Ramsden, 1992). Each of these qualitatively different learning approaches is based on different learning motives and has an associated preferred learning strategy. Deep approaches to learning are motivated by intrinsic interest in the subject area, and the intention is to understand and satisfy interest by looking for meaning, connections between ideas, and relationships to previous knowledge (Biggs, 1987). In contrast, surface approaches are based on extrinsic motivation, and the intention is instrumental—to "get by" with minimal effort by rote learning essential information to reproduce it on examinations.

Deep learning approaches are consistently associated with better-quality learning outcomes (Prosser and Trigwell, 1999; Zeegers, 2001), while surface approaches to learning result in student dissatisfaction and poor educational outcomes:

> Surface approaches have nothing to do with wisdom and everything to do with aimless accumulation. They belong to an artificial world of learning, where faithfully reproducing fragments of torpid knowledge to please teachers and pass examinations has replaced understanding. "Paralysis of thought" leads inevitably to the misunderstandings of important principles, weak long-term recall of detail, and inability to apply academic knowledge to the real world. A surface approach . . . leads down the same desolate road in every field. (Ramsden, 1992, p. 60)

Inappropriate assessment regimes direct students down this "desolate road" by rewarding or pressuring them into ineffective surface learning approaches. Surface learning is associated with assessment that establishes excessively heavy workloads (possibly in an attempt to coerce students into learning), or overemphasizes recall of factual information to the detriment of higher-order thinking skills. In addition, anxiety-provoking assessments, lack of student choice in assessment, and assessment of isolated parts of the curriculum rather than integrated assessment tasks mitigate against deep approaches to learning (Ramsden, 1992).

ASSESSMENT STRATEGIES FOR FIRST-YEAR SCIENCE

Multiple challenges surround the issue of assessment in resource-deprived, large, diverse first-year classes. Assessment needs to be used strategically to enhance learning by focusing student attention on the range of learning objectives and discouraging surface approaches to learning. Deep learning approaches and better learning outcomes are fostered by using a variety of assessment types to minimize disadvantage to particular groups of students (Brown and Knight,

1994), by incorporating some choice into formal assessment tasks, and by making these choices professionally relevant to the subgroups of students (Ramsden, 1992) that are particularly diverse in broad introductory units.

Assessment techniques should be based on criteria made explicit to students before the task is undertaken, provide motivating and challenging authentic tasks relevant to students' interests, and supply students with frequent feedback to enhance learning. The need to provide students with high-quality formative assessment and feedback cannot be overstated, and this is "more important during the initial period of transition into university than at any other time" (McInnes, James, and McNaught, 1995, p. 123). One option especially relevant to first year, though not particular to science students, is providing a formal, low-stakes written assessment task early in the first semester (James, McInnis, and Devlin, 2002). This can be used to encourage successful students, as well as to identify students at potential risk of failure and/or with literacy problems, who may then be referred to appropriate support services for further formative assessment and assistance.

SPECIFIC ASSESSMENT TECHNIQUES FOR LARGE FIRST-YEAR SCIENCE CLASSES

Some formal assessment techniques that have been used successfully in the context of introductory science classes are described next. This array of techniques is by no means definitive but does serve to sample the variety of alternatives available. Most of these techniques can be used for a combination of formative and summative purposes, or purely for formative feedback to students. Advantages and disadvantages of these techniques are described, along with some suggestions to increase assessment efficiency by reducing staff workload or maximizing the learning potential of resource-intensive assessments, such as written assignments. Examples of successful use in introductory science classes are given, and in most cases more information can be found in the original source. The assessment techniques are organized along an approximate continuum from easily administered summative assessments, such as examinations, to assessments that are more resource intensive and formative. The discussion ends with consideration of self- and peer assessment, which can be applied to most specific assessment categories mentioned and assessment of group work.

Examinations

The short-answer end-of-semester examinations conventionally used in first year as a reliable and "efficient" measure of declarative scientific knowledge have been criticized on a number of fronts. They are inauthentic, anxiety-inducing, unintegrated, inappropriate for project-based work, and often have little formative assessment value. Strong concerns exist about an inappropriate narrow focus on lower- order thinking skills (Ramsden, 1992). For these reasons, high-stakes, summative end-of-semester examinations may instill ineffective surface learning approaches, may hamper meaningful learning, and may result in distorted judgment and reporting of students' capabilities and understanding. They should therefore be used only "very cautiously, preferably in combination with other methods" (Ramsden, 1992, p. 211).

If examinations or similar tests are used cautiously, with a focus on formative assessment, it can be argued that they may have a role in learner-centered education. It is claimed that when well constructed and used appropriately they do have value as valid tests of core content knowledge (Hughes and Magin, 1996) and can test higher-order cognitive skills such as analysis and synthesis (Brown and Knight, 1994). There are guidelines on constructing and validating objective tests in Hughes and Magin (1996), along with examples of different examination question types relevant to science. Using common scientific misconceptions as distractors and asking for best rather than correct answers is thought to enhance the utility of multiple-choice questions in science (Tamir, 1996). Making past exam papers available aids learning, as do open-book or take-home exams. Nonetheless, it is clear that valid and authentic assessment of students' abilities requires assessment techniques other than examinations.

Self-Marked Quizzes

Self-marked objective tests can be used formatively throughout the semester by providing students with sets of short-answer or multiple-choice questions, together with the answers and brief explanatory feedback notes (Ramsden, 1992). This technique helps students to identify misconceptions or problematic content areas themselves, early enough for remedial action. Paper-based or online versions of these are relatively commonly reported in the literature (e.g., Peat, 2000).

Online and ICT-Based Assessment

Currently available information and communication technologies (ICTs) such as WebCT are providing increasing opportunities for automated, immediate feedback on students' conceptual understanding at quite sophisticated levels (James, McInnis, and Devlin, 2002; Schwartz and Webb, 2002). Computer-assisted assessment of virtual inquiry-based science is particularly suited to large classes and off-campus students and is commonplace in the UK Open University (Laurillard, 1998). Online assessment tasks can include answering short-answer questions, labeling diagrams, doing calculations, manipulating graphs, running simulations, and the like, and the best ICT-based assessment uses a range of these tasks rather than just multiple-choice questions (Brown, 2002). Reported experience in the first-year science context suggests that online resources are unlikely to be used unless they are linked with formal summative assessment in some way (e.g., Green, 1992).

Use of summative online quizzes has been reported in helping students to develop calculation skills in first-year chemistry (Schwartz and Webb, 2002), and more broadly in first-year biology and geology (Peat, 2000; Peat and Hewitt, 1998). Use of computer-based summative examinations has been described for biology (e.g., Zakrzewski and Bull, 1998) and physiology and pharmacology (Dallemagne in Hughes and Magin, 1996) with reported gains in efficiency once implemented and favorable student and staff response. Computer-based summative examinations have similar validity and authenticity problems as paper-based examinations; however, as argued by Zakrzewski and Bull (1998) if they result in cost savings allocated to other more resource-intensive formative assessment tasks, overall learning may be enhanced.

Assessment Techniques for Practical Laboratory Work

Laboratory sessions are central to most tertiary science courses, however, they are usually staffed by inexperienced, poorly paid, and undersupported casual staff (McInnes, James, and McNaught, 1995). Formative assessment of laboratory skills and conceptual understanding can therefore be enhanced by providing demonstrators with professional development in assessment and feedback techniques, along with detailed marking criteria, marking guides, and standardized feedback sheets for assessments (James, McInnis, and Devlin, 2002). A positive tone to feedback, especially critical feedback, is paramount, and it is vital that demonstrators have this skill.

A range of assessment techniques is available to assess basic procedural skills such as ability to cut hand sections or use a microscope. Simple tick-a-box checklists can be used quickly by demonstrators or by the students themselves, to assess performance in procedural skills (e.g., Tamir, 1996). Mastery learning, a type of criterion-referenced assessment, is another useful technique in assessing scientific procedures (Dunn, 1986). In mastery learning, students are provided with a set of specific criteria related to a particular skill; they practice the skill at their own pace until they consider it mastered, then, when they feel ready, take a criterion-referenced test. If assessed as satisfactory they proceed to the next skill, otherwise they repeat the assessment. A sample set of criteria for mastery learning of microscope technique, following an example described by Dunn (1986), might include the following points:

- Set up the microscope for maximum image quality.
- Focus the microscope on a slide using all three objectives.
- Estimate the diameter of the image on the slide.
- Draw a diagram of the image.
- Draw a scale bar to indicate the size of the image.

As an alternative to commonly used postlaboratory tests, short summative prelaboratory quizzes have been used to encourage students to prepare for and get the most value out of laboratory sessions, for example, in first-year chemistry (Tronson, 2002, in James, McInnis, and Devlin, 2002).

Assessment of Laboratory Reports, Notebooks, and Journals

Laboratory reports are frequently used to foster and assess student understanding of scientific content and procedures, as well as initiating students into the specific conventions of scientific writing. Writing reports during the relevant practical session(s) rather than later has the advantages of minimizing plagiarism, motivating some students, and being shorter and less time consuming for both students and markers. However, realistic time must be available to the students or this may provoke undue stress, which is counterproductive to meaningful learning. If it is considered important that several reports are written up, marking of these can be selective, based on student choice or as a random selection by the marker. In an illuminating account of transforming a large introductory physics class, Green (1992) halved the number of laboratory reports to be submitted after noting little apparent improvement past the first three or four reports. Alternatives to laboratory reports are structured learning journals or logs documenting scientific procedures and thinking (Gibbs, Habeshaw, and Habeshaw, 1988). Laboratory notebooks are a valuable tool for fostering development of laboratory skills and can be used for informal or formal assessment purposes or as bases for discussion of scientific processes.

Assessment of Attitudinal Factors

The question of assessing personal student attitudes is known to be problematic (Clanchy and Ballard, 1995; Rowntree, 1981). Some of the issues associated with assessment of attitudes in the context of undergraduate medical education are described by Toohey (2002), who concludes that attitudinal assessment is both desirable and feasible but should be conducted formatively rather than summatively. Radloff and de la Harpe (2001) also advocate formative assessment of a number of attitudes and describe some generic questionnaires for the purpose. There are a number of questionnaires available to investigate student critical thinking, curiosity, and interest in a science context, described by Tamir (1996).

Demonstrated unethical conduct such as plagiarism can readily be assessed. James, McInnis, and Devlin (2002) provide a comprehensive list of specific suggestions to help deal with plagiarism, aimed at making expectations clear to students, minimizing plagiarism opportunities, and reacting visibly and strongly to plagiarism. Some of the specific suggestions that are particularly suitable for large first-year classes include teaching students about summarizing, paraphrasing, referencing, and citation conventions to minimize inadvertent plagiarism. Opportunities for deliberate plagiarism can be minimized by designing assessment tasks that are highly specific to the particular subject matter, unique from year to year, or that integrate personal reflection or experience. Students can also be asked to retain evidence that they have not plagiarized, such as essay plans, photocopied articles, or library call numbers. Students can also be made aware of the available electronic plagiarism detection devices, and the ease with which specific phrases in their work can be detected using standard Internet search engines.

Assessment of Written Assignments

Individual essays, assignments, and reports are valuable means of assessing skills of written communication and information literacy, as well as understanding of content. Written assignments, though, are highly resource-intensive, both in terms of student and staff time and effort. It is therefore becoming more common for subject specialists to maximize the learning potential of these assessments by developing them in cooperation with academic and/or language skills specialists. This cooperative approach is based on the fact that a sizable proportion of current first-year students need specialist help with English literacy skills, particularly some students from non-English-speaking backgrounds. Subject-specific assessment tasks are the most effective context for students to develop study and literacy skills, as generic assistance is not as effective (for reviews see Hadwin and Winne, 1996; Hattie, Biggs, and Purdie, 1996). Some increasingly common strategies for written assessments include subject specialists developing the assessment questions, instructions and marking criteria in conjunction with academic skills staff, team teaching of context-specific essay- or report-writing workshops, and allowing resubmission of failed assignments if students seek advice from study skills specialists. Specialist literacy support linked to assessment tasks is increasingly being provided online, in conjunction with virtual tutoring and online discussion forums to enhance student activity, motivation, and feedback associated with formal written assessments.

The learning potential of written assessment tasks can also be enhanced by asking students to assess model essays or reports and exemplars of different quality assignments. This can help students to identify the conventions and expectations of the disciplinary discourse (Brown and Knight, 1994; James, McInnis, and Devlin, 2002), especially in their first written assignment.

An efficient way to assess written assignments is by using standardized feedback sheets. These incorporate a numbered list of common errors, together with associated explanations and further reading, and are returned to the students cross-linked to their work by a numbered code written where relevant on assignments. This can expedite rapid high-quality feedback on assignments, and allow for much more detailed feedback per student than might otherwise be the case (Ramsden, 1992). These do not preclude the marker writing spontaneous feedback on assignments, but do save a great deal of time in providing feedback about more common problems.

Even using this technique, full essays or reports are time-consuming for staff to mark (and students to write). Alternative shorter assessment items could include summaries or paraphrases based on some relevant article, a letter to a friend explaining in simple terms some scientific concept, or partial essays/reports (Brown and Knight, 1994). Concept maps are an alternative method of assessing student understanding of scientific factual knowledge and are particularly useful to encourage and measure students' relational understanding—how scientific concepts fit together (Novak, 1996). Concept maps can be produced and marked relatively quickly, and they may also be used in conjunction with written assignments.

Self- and Peer Assessment

Self-assessment, according to Boud (1986), involves students in identifying criteria by which to measure their own work and assessing how well they have met these criteria. Peer assessment involves groups of students rating the work of their peers. These assessment techniques are effective tools for improving the metalearning of students (for a review see Dochy, Segers, and Sluijmans, 1999), many of whom in their first year still need to learn how to learn. By focusing students' attention on the strengths and weaknesses in sample work and work of their peers, students gain a new perspective on their own work, learning to internalize appropriate standards and "calibrate" the effort required in their own work (Gibbs, 1999). These processes can promote students' ability to monitor and judge their own learning, foster student reflection on their learning, and increase student confidence and independence (Boud, 1986), as well as enhance motivation and deep learning (Somervell, 1993, in Dochy, Segers, and Sluijmans, 1999).

Although appealing on these grounds, as well as for their potential to ameliorate staff marking loads, the use of these strategies in large first-year classes requires careful consideration. Establishing such alternative assessment regimes may initially be more time-consuming for staff (Brown and Knight, 1994). Other potential disadvantages to peer assessment include possible student hostility, undesirable comments in formative feedback, and reluctance or overenthusiasm to criticize peers. A climate of trust and respect is therefore important. The jury is still out about the reliability and validity of self- or peer assessed marks in summative tasks; for example, in a self-assessed first-year biology context there was overmarking of poster presentations by "poor" students and undermarking by "good" students (Orsmond, Merry, and Reiling, 1997). However, in an action research study into peer assessment in large classes, Ballantyne, Hughes, and Mylonas (2002) conclude that the learning advantages of this form of assessment outweigh the difficulties. They provide a list of recommendations for implementing peer assessment in large classes, summarized here (modified from Ballantyne, Hughes, and Mylonas, 2002 pp. 436-437):

- When using with first-year students, structure peer assessment processes carefully.
- Task should be stand-alone, familiar to staff and students, and require critical analysis by assessors.
- Instructors (and demonstrators) should be in agreement and the process should be discussed fully in a staff workshop prior to implementation, with all staff clear about the process and their role.
- Staff-generated assessment criteria should be used and the process explained to students clearly and persuasively at a lecture.
- Using marks summatively enhances student commitment to the task.
- Students should be trained in practice sessions using exemplars of variable quality, and receive feedback.

- Preserve anonymity of assignments and markers.
- Using staff to moderate the process, by double marking a small sample of assignments, would enhance perceived fairness and allow for monitoring of student learning.
- Allow for re-marks by instructors where written justification is supplied.
- Limit the use of peer assessment across the academic program.

Several published examples of the successful use of self- and peer assessment in introductory science classes are available. They have been used in formative assessment of intermediate stages of assignments in health and nursing classes (Scoufis, 1999), in a problem-based learning approach to first-year computer science (Peat and Hewitt, 1998), in oral presentations in pharmacology (Brown and Knight, 1994), and in assessment of student posters in biology (Garvin et al., 1995, as cited in Bourner, Hughes, and Bourner, 2001). In one case study, formative peer assessment of problem sheets in a compulsory second-year engineering class was associated with markedly better performance in the examination. This was thought to be due to increased time on task and learning activity, almost immediate feedback after submission, and the social pressure of being assessed by peers (Gibbs, 1999). Student self-assessment can be a requirement of any submitted work, using criterion-based proformas that are filled out by students and submitted with assignments (Brown and Knight, 1994). An excerpt of a sample self-assessment proforma is shown in Box 10.1, and this could be used formatively by students and summatively by staff.

Assessment of Group Work

Group project-based learning is common in more inquiry-oriented and student-centered science curricula (Fraser and Deane, 1998). For large first-year classes, group work may be shorter and more structured than a full project but ideally is more explorative than laboratory exercises (Hegarty-Hazel, Boud, and Dunn, 1987). Group projects bring with them opportunities and imperatives for novel, authentic assessment that is integrated with the particular project or problem situation. An example of such authentic assessment is described in the context of second-year biostatistics by Panizzon and Boulton (see

BOX 10.1. Example of Self-Assessment Pro Forma for Formative Use by Students

A Guide to Help You Assess the Quality of Your Own Work

Read the descriptions against each of the following criteria carefully, and circle the grade that you consider applies to each aspect of your work. When the assignment is returned, compare your assessment with that of your marker to identify your problem areas.

Introduction: 15 percent

N   Superficial discussion of the background issues, aims not stated

P   Limited discussion of some relevant background issues, aims stated unclearly

Cr Adequate discussion of most background issues, aims stated clearly

D   Thorough discussion of most relevant background issues, aims stated clearly

HD Discussion of background issues covers all relevant aspects in a coherent, integrated fashion, leading to and justifying a clear statement of aims

Results: 30 percent

N   Incomplete or indecipherable

P   All results presented and decipherable, include several errors or mistakes in presentation, such as repetition of visually presented results in text. Includes material misplaced in results section, such as discussion of previous research.

Cr Results with few errors, presented mostly according to scientific convention as outlined in unit handbook. Visual results summarized briefly in text. No extraneous discussion in results section.

D   Results mostly accurate and presented in accordance with scientific convention as instructed. Visual results presented clearly, summarized coherently and briefly in text.

HD Results presented extremely accurately and clearly, all conventions followed. Communication of results in text and graphically is particularly comprehensive, clear, and concisely presented.

Chapter 8). As pointed out by James, McInnis, and Devlin (2002), assessment is crucial to the success or otherwise of group work. Summative grades for group work can be applied to the entire group, which potentially reduces marking loads on staff if handled appropriately. However, both the product and the process of group work need to be assessed, and assessment of the group work process often requires some form of peer and/or self-assessment of contribution among group members, possibly moderated by staff (Brown and Knight, 1994). Techniques for taking into account unequal contributions by members include adding an individual contribution factor to the group grade or the group distributing the shared mark to its members on the basis of their relative contributions. Further guidelines for assessing group work are provided by James, McInnis, and Devlin (2002).

Examples of successful group assessment include peer-assessed poster presentations of short group projects in first-year biology (Garvin et al. as cited in Bourner, Hughes, and Bourner, 2001), and group work accounting for 22 percent of the semester mark in a human biology service course of about 600 students, both on- and off-campus (Fyfe, 2000). Group projects are summatively assessed in an introductory zoology class (Jones, 2002, in James, McInnis, and Devlin, 2002), emphasizing the product and giving the same mark to all members of the group, but with an adjustment if necessary according to students' assessment of their contributions to the product.

CONCLUSION

The central role of assessment in reforming tertiary education is receiving growing recognition from tertiary educators and commentators (see, e.g., Brown and Knight, 1994; James, McInnis, and Devlin, 2002). The argument central to this chapter is that assessment-driven reform is needed above all in the first year of study at university, which is so critical for individual student and institutional success. Both the problems and potential benefits of assessment are magnified in large classes of diverse first-year students, who are learning what is valued in tertiary education. Assessment choices in first year must therefore be centered on the learning needs of students; by genuinely reflecting learning objectives, by avoiding rewards for reproduction of superficially understood, rote-learned information, and by emphasizing formative assessment and feedback. There are a multitude of options for formally assessing students, and many of these are possible even by time-pressured staff in resource-limited large classes.

One of the emerging trends in assessment of large first-year classes, particularly in flexible learning situations, is the use of ICTs to combine assessment with learning in increasingly powerful and sophisticated ways. In addition, assessment of group work is becoming more common as inquiry-based learning is filtering down to the first-year level. Finally, use of self- and peer assessment is reflecting growing awareness of the importance of skills of self-evaluation and independent autonomous learning. Group, self-, and peer assessment are still relatively infrequently used at the first-year level, despite documented advantages for learning, and are therefore potentially fruitful avenues for further research. Action research describing implementation of these assessment techniques within clearly described contexts would greatly assist teaching staff who are considering adopting some of these alternatives.

The examples presented here demonstrate that innovative assessment methods can be successfully and efficiently used in large first-year classes, although some may require an initial heavy input of staff resources and planning. It is also apparent that small changes associated with more conventional assessments in science can maximize student learning and marking efficiency. The inertia of established procedures, the potential risks of change in large classes, and resource limitations all tend to mitigate against innovation in first-year assessment, precisely where it is most needed. It is hoped that the examples presented here will demonstrate the range of assessment alternatives that can be adopted to foster learning in the crucial first year of tertiary study.

REFERENCES

Ballantyne, R., Hughes, K., and Mylonas, A. (2002). Developing procedures for implementing peer assessment in large classes using an action research process. *Assessment and Evaluation in Higher Education,* 27(5): 427-441.

Biggs, J. (1987). *Student approaches to learning and studying.* Melbourne: Australian Council for Educational Research.

Boud, D. (1986). *Implementing student self-assessment* (Green Guide Number 5). Kensington: Higher Education Research and Development Society of Australasia.

Bourner, J., Hughes, M., and Bourner, T. (2001). First-year undergraduate experiences of group project work. *Assessment and Evaluation in Higher Education,* 26(1): 19-39.

Brown, S. (1999). Institutional strategies for assessment. In S. Brown and A. Glasner (Eds.), *Assessment Matters in Higher Education.* Buckingham, UK: The Society for Research into Higher Education and Open University Press.

Brown, S. (2002). What to do about John? In Schwartz, P. and Webb, G. (Eds.), *Assessment: Case studies, experience and practice from higher education* (pp. 32-38). London: Kogan Page.

Brown, S. and Knight, P. (1994). *Assessing learners in higher education.* London: Kogan Page.

Clanchy, J. and Ballard, B. (1995). Generic skills in the context of higher education. *Higher Education Research and Development,* 14(2): 155-166.

Dochy, F., Segers, M., and Sluijmans, D. (1999). The use of self-, peer, and co-assessment in education: A review. *Studies in Higher Education,* 24(3): 331-350.

Dunbar, R. (1995). *The trouble with science.* London: Faber and Faber.

Dunn, J. (1986). Assessment of students. In D. Boud, J. Dunn, and E. Hegarty-Hazel (Eds.), *Teaching in laboratories* (pp. 79-106). Guildford, England: Society for Research into Higher Education and NFER-Nelson.

Entwistle, N. and Marton, F. (1984). Changing conceptions of learning and research. In F. Marton, D. Hounsell, and N. Entwistle (Eds.), *The experience of learning* (pp. 211-228). Edinburgh: Scottish Academic Press.

Fraser, S. and Deane, E. (1998). *Doers and thinkers: An investigation of the use of open learning strategies to develop life-long learning competencies in undergraduate science.* Canberra, Australia: Department of Employment, Education, Training and Youth Affairs.

Fyfe, G. (2000). Active learning for understanding in introductory human biology. Paper presented at the Teaching and Learning Forum, Perth, Australia.

Gibbs, G. (1999). Using assessment strategically to change the way students learn. In S. Brown and A. Glasner (Eds.), *Assessment matters in higher education* (pp. 41-53). Buckingham, UK: The Society for Research into Higher Education and Open University Press.

Gibbs, G., Habeshaw, S., and Habeshaw, T. (1988). *53 interesting ways to assess your students.* Bristol, United Kingdom: Technical and Educational Services Ltd.

Green, A. (1992). Teaching introductory physics: Techniques and resources for large classes. In G. Gibbs and A. Jenkins (Eds.), *Teaching large classes in higher education: How to maintain quality with reduced resources* (pp. 99-116). London: Kogan Page.

Hadwin, A. and Winne, P. (1996). Study strategies have meagre support: A review with recommendations for implementation. *Journal of Higher Education,* 67: 692-715.

Hattie, J., Biggs, J., and Purdie, N. (1996). Effects of learning skills interventions on student learning: A meta-analysis. *Review of Educational Research,* 66: 99-136.

Hegarty-Hazel, E. (Ed.) (1990). *The student laboratory and the science curriculum.* London: Routledge.

Hegarty-Hazel, E., Boud, D., and Dunn, J. (1987). Strategies for learning scientific skills in the laboratory. In A. Miller and G. Sachse-Akerlind (Eds.), *The learner in higher education: A forgotten species?* (pp. 51-57). Sydney: Higher Education Research and Development Society of Australasia.

James, R., McInnis, C., and Devlin, M. (2002). *Assessing learning in Australian universities.* <www.cshe.unimelb.edu.au/assessinglearning>.

Laurillard, D. (1998). Computers and the emancipation of students: Giving control to the learner. In P. Ramsden (Ed.), *Improving learning: New perspectives* (pp. 215-233). London: Kogan Page.

Laws, P. (1996). Undergraduate science education: A review of research. *Studies in Science Education,* 28: 1-85.

Lowe, I. (1994). Custom-built irrelevance: The problem of traditional universities. *Research and Development in Higher Education,* 16: 1-7.

McInnes, C., James, R., and Hartley, R. (2000). *Trends in the first-year experience.* Canberra, Australia: Department of Education, Training and Youth Affairs.

McInnes, C., James, R., and McNaught, C. (1995). *First year on campus.* Melbourne, Australia: Centre for the Study of Higher Education, University of Melbourne.

Niland, J. (1998). *The fate of Australian Science—The future of Australian universities.* <http://www.avcc.edu.au/news/public_statements/ speeches/1998/fate.htm>.

Novak, J. (1996). Concept mapping: A tool for improving science teaching and learning. In D. Treagust, R. Duit, and B. Fraser (Eds.), *Improving teaching and learning in science and mathematics* (pp. 32-43). New York: Teachers College Press.

Orsmond, P., Merry, S., and Reiling, K. (1997). A study in self-assessment: Tutor and students' perceptions of performance criteria. *Assessment and Evaluation in Higher Education,* 22(4): 357-370.

Passey, R.H. (1996). The curriculum—A change in emphasis. In J.S. Ryan (Ed.), *Rural science: Philosophy and application* (pp. 176-200). Armidale, Australia: School of Rural Science, University of New England.

Peat, M. (2000). Online assessment: The use of Web-based self-assessment materials to support self-directed learning. Paper presented at the Teaching and Learning Forum, Perth, Australia, February 2-4.

Peat, M. and Hewitt, R. (1998). Improving the first-year experience: To set a new culture in place, faculty of science style. Paper presented at the Third Pacific Rim Conference on the First Year in Higher Education: Strategies for Success in Transition years, Auckland, New Zealand, July 5-8.

Pitkethly, A. and Prosser, M. (2001). The First Year Experience Project: A model for university-wide change. *Higher Education Research and Development,* 20(2): 185-198.

Prosser, M. and Trigwell, K. (1999). *Understanding learning and teaching: The experience in higher education.* Buckingham, UK: The Society for Research into Higher Education and Open University Press.

Radloff, A. and de la Harpe, B. (2001). Expanding what and how we teach: Going beyond the content. Paper presented at the Teaching and Learning Forum, Perth, Australia, February 7-9.

Ramsden, P. (1992). *Learning to teach in higher education.* London: Routledge.

Rowntree, D. (1981). *Developing courses for students.* London: McGraw-Hill.

Rowntree, D. (1987). *Assessing students: How shall we know them?* London: Kogan Page.

Schwartz, P. and Webb, G. (2002). *Assessment: Case studies, experience, and practice from higher education.* London: Kogan Page.

Scoufis, M. (1999). How not to burden first-year students and lecturers with multiple assessment tasks and yet still measure desired learning outcomes? In K. Martin, N. Stanely, and N. Davison (Eds.) *Teaching in the Disciplines/Learning in Context* (pp. 373-377). Proceedings of the 8th Annual Teaching Learning Forum. The University of Western Australia, February 1999. Perth: UWA. <http://lsn.curtin.edu.au/tlf/tlf1999/scoufis.html>.

Stokstad, E. (2001). Information overload hampers biology reform. *Science,* 293: 1609.

Tamir, P. (1996). Science assessment. In M. Birenbaum and F. Dochy (Eds.), *Alternatives in assessment of achievements, learning processes, and prior knowledge* (pp. 93-130). Boston: Kluwer Academic Publishers.

Toohey, S. (2002). Assessment of students' personal development as part of preparation for professional work—Is it desirable and is it feasible? *Assessment and Evaluation in Higher Education,* 27(6): 529-538.

Tronson, D. (2002). Assessing pre-laboratory work: Is what's good for the goose necessarily good for the gander? Case study in R. James, C. McInnis, and M. Devlin (Eds.), *Assessing Learning in Australian Universities.* Melbourne: Centre for the Study of Higher Education and The Australian Universities Teaching Committee. <http://www.cshe.unimelb.edu.au/assessinglearning/04/case15.html>.

Yorke, M. (1998). The management of assessment in higher education. *Assessment and Evaluation in Higher Education,* 23(2).

Yorke, M. (2001). Formative assessment and its relevance to retention. *Higher Education Research and Development,* 20(2): 115-126.

Zakrzewski, S. and Bull, J. (1998). The mass implementation and evaluation of computer-based assessments. *Assessment and Evaluation in Higher Education,* 23(2): 141-152.

Zeegers, P. (2001). Approaches to learning in science: A longitudinal study. *British Journal of Educational Psychology,* 71: 115-132.

Chapter 11

Exploring the Usefulness of Broadband Videoconferencing for Student-Centered Distance Learning in Tertiary Science

Robyn Smyth

INTRODUCTION

In recent years, the face of higher education in Australia and internationally has begun to change from its traditional form. First, students with a wider range of backgrounds and abilities are now entering into higher education and many more of them study in the distance mode (off campus). Second, economic rationalism has become the driver for government policy with its focus on value for money and greater accountability. Third, educators have begun to examine teaching and learning practices and processes from a more holistic viewpoint in an effort to improve learning outcomes (Panizzon, Pegg, and Mulquiney, 1999). These trends are mirrored in other parts of the world, particularly in the United Kingdom (Quality Assurance Agency, 2000). The conceptual framework for this chapter accepts the first two changes and acknowledges their strong influence in a contextual sense, while concentrating on the challenges they provide and beginning to explore effective use of broadband technology for mixed-mode distance education and student-centered teaching and learning in tertiary science.

This chapter, therefore, will explore the means by which traditional approaches to teaching in the sciences may be enhanced by integrating one of the emerging information and communication technologies (ICT), Internet-based videoconferencing, using medium to large room systems. It will explore these issues in the context of managing changes from teacher-centered to student-centered teaching and learning.

A CONCEPTUAL FRAMEWORK: PEDAGOGY, PHILOSOPHY, AND TRANSITIONS

Considerations of pedagogy (the way we teach) and philosophy (they way we view knowledge acquisition) are not inconsiderable when it comes to discussing improving outcomes from educative processes, and they deserve longer treatment than is possible here. In terms of pedagogy, student-centered approaches to teaching and learning in higher education are emerging as more suitable strategies to promote deeper learning (Trigwell and Prosser, 1999). A student-centered approach is a pedagogical style underpinned by philosophical or conceptual views that place learning rather than teaching at the core of teaching practice (Ramsden, 1992; von Glasersfeld, 1995; Fosnot, 1996; Foley, 2000). The motivation for promoting such an apparently radical shift away from traditional practice stems from growing realizations that the actions of learners have a more important influence on the outcomes learners achieve than the actions of teachers (Biggs, 1999). A consequence of the shift has been the increasing interest in more flexible delivery of learning utilizing ICT, including broadband Internet-based videoconferencing. The balance between improving learning outcomes and the cost of doing so will impact most significantly on the quality of ICT as pedagogical tools (Kirkpatrick and Jakupec, 1999; Quality Assurance Agency, 2000; Twigg, 2002).

As the use of ICT gathers momentum, teachers' and students' views about how and what to teach and their changing roles in the educative process are being challenged. For example, what new skills do teachers need to support their use of ICT in teaching? Managing the transition from traditional settings and teaching approaches, roles, and responsibilities to new ones associated with a more student-centered focus and the implementation of ICT will become a key factor to successful migration from didactic to learner-centered practice. In attempting to shed insight into these challenges, the following considerations are addressed:

- The types of changes needed in teaching approaches
- The impact of changes that can challenge stakeholders' belief systems about education

- The importance of planning to manage transitions from existing to future practice
- The potential benefits of student-centered approaches for teaching science utilizing new technologies

Changing Pedagogical Frameworks

Two clarifications of meaning should be made clear at this point. First, the term *pedagogy,* defined earlier as *the way we teach,* is intended to represent evolving practice rather than fixed ideals, since teaching practice is most generally regarded as action characterized by continuous subtle change (Fullan, 1991). Second, changing pedagogy is acknowledged to be a complex process which reaches into the soul of most teachers and can be affronting to their deeply held beliefs about the way they teach (Hargreaves, 1998). Leveraging the deep change required to make a radical shift in many teachers' pedagogies requires that they be given time and support to undertake thoughtful planning (Boyer, 1990; Hargreaves, 1997b,c; Fullan, 1998; Hargreaves et al., 1998; Louis, Toole, and Hargreaves, 1999).

Most successful teachers tend to be good adapters. They are able to take new and emerging practices and strategies and integrate them into their teaching in seamless ways that make them easily acceptable to students (Fullan, 1991). For example, the use of e-mail, mailing lists, and voicemail instead of letters mailed out has been adopted as a means of reducing the isolation of distance-education students through more frequent and rapid communication (Salmon, 1998; Wylie, 1998). For many teachers, the alteration of practice in this manner is more common than deeper reflection on their primary motivation for teaching the way that they do, simply because they are not given sufficient time to reflect, move forward, reflect and move forward again (Adelman and Walking Eagle, 1997). In these times of rapid change, it is becoming increasingly important for good teachers to understand why they do what they do well so that they can select appropriate practices and pedagogies to enhance their expertise. For teachers in higher education, keeping up to date with discipline-based knowledge, practice, and understanding is one part of enhancing their teaching. Keeping abreast of fundamental shifts in educators' understanding about how learning takes place is another. This knowledge lies at the heart of improving the outcomes of students' learning (Biggs and Collis, 1982).

Until late last century in Western countries, much educational theorizing was inherited from the psychological sciences, which focused on learning in terms of computational cognitive models. The recognition that knowledge may be constructed by individuals as they go about their daily living and the increasing interest in the *situated* nature of learning has refocused thinking about education toward knowledge as an aggregation of social experiences rather than an aggregation of imparted wisdom (Grundy, 1987; Gibbs, 1995; Laurillard, 2002). From this perspective, the role of the teacher changes from that of knowledge transmitter to that of learning facilitator, while the type of teaching strategies become less frequently didactic and teacher dependent to more frequently collaborative and student generated (Mitchell and Hope, 2002). Equally important has been the recognition that teaching is not only a profession but also an emotional vocation (Hargreaves, 1998). Much of the creative energy that teachers bring to their work is derived from their beliefs and feelings about what they do.

Effecting Changes in Pedagogy

Through reflection based on scholarship, individual teachers evaluate, trial and introduce new ways of teaching into their practice. In the case of changing practice from teacher- to student-centered, a shift in practice involves much more than the adaptation of new strategies (Buckley, 2002). A fundamental shift in the way the teacher views his or her purpose for teaching is required, because this change represents a conceptual shift away from focusing on the teacher's needs to focusing on those of the students (Mezirow, 1991). As a consequence, the source of the teacher's personal satisfaction may be challenged. For some teachers, the challenge is a threat to their perceptions of the power relationship between themselves and their students. Equally, students should be well prepared for changed practice since they also have learned conceptions and preferred roles in the learning environment, which may be equally challenged (Smyth, in press).

The depth of scholarship is important here. To effect significant changes in pedagogy and practice, teachers need to investigate, discuss, trial, and reflect upon new approaches and strategies thoroughly (Boyer, 1990; Webb, 1996). Endeavoring to alter pedagogy should encompass

- research of recent literature and consideration of appropriate pedagogical frameworks (Andresen, 2000);
- trials in circumstances where there is no risk of untoward consequences for either teachers or students;
- modeling for feedback (demonstrating successful and relevant techniques and seeking feedback from students and colleagues);
- seeking pedagogical expertise from educators when necessary; and
- planning and managing change to increase opportunities for success and seeking positive feedback aimed at promoting continuous improvement (Mezirow, 1991; Trigwell et al., 2000).

These techniques are premised on promoting changed conceptions of teaching and learning as well as changed beliefs about knowledge acquisition. Since fundamental beliefs can either hinder or promote openness toward deep change, they require brief consideration here.

Investigating and Changing Philosophical Frameworks

In many cases, teachers may be in the process of adopting student-centered approaches to teaching and learning in environments where didactic or teacher-centered approaches have been the norm. Articulation of the conceptions of both teachers and students concerning their roles in the teaching and learning process should be acknowledged and discussed (Smyth, 2002).

Discussion of lecturers' and students' beliefs about learning can provide opportunities to explore values and attitudes, while the challenges or fears that are faced by both stakeholders can be identified. Action research is an appropriate methodology for engaging teachers in meaningful discussion about changing their teaching, as it fosters a view of change as a continuous process of improvement and its philosophical underpinnings are constructivist (Carr and Kemmis, 1986). The centrality of feedback and reflection in the action research approach assists them to go beyond routine evaluation of teaching toward deeper understanding (Ramsden, 1992; Gibbs, 1995; Trigwell et al., 2000). Appropriately used, the action research cycle involves participants in ongoing learning as they investigate issues in practice, think, discuss and plan how improvements might be undertaken, then observe and reflect upon the outcomes and undertake more planning for improvement (Gibbs, 1995).

Engaging students in the process of action research, discussing pedagogical change, and encouraging them to provide feedback has the potential to reduce resistance to change, as students will have the opportunity to express their views. In addition to enhancing the action research process, an appreciation of the philosophical foundations of knowledge acquisition can provide useful insight about personal attitudes and values that promote or hinder change (Hannah, 1979; Grundy, 1987, 1992; Mezirow, 1991).

Making Transitions

It is worth emphasizing that changing the practices of teaching and learning should not be regarded in isolation from the context of the broader disciplinary and institutional environments (Fullan, 1997a; Hargreaves, 1997a; Laurillard, 2002).

Consideration of the educational change literature provides core strategies that can underpin successful change on micro and macro scales. On a micro level, individuals may consider an action research approach to making changes in practice (Gibbs, 1995). On a macro level, appropriate strategies include vision building, evolutionary planning, leadership, and engaging participants intellectually and emotionally (Hargreaves, 1996; Fullan, 1997b; Cooksey, 2000).

Moving toward conceptions of teaching that promote discourse and collaborative learning while developing respect for teachers as learners and learners as teachers should foster trust through developing a sense of community about change (Mezirow, 1990, 1991). Showing explicitly how the contributions of both teachers and students will be valued and included in the dialogue also encourages a sense of community and engages it as a source of positive emotion and hope (Sergiovanni, 1998). Including time and space for incorporating and trying out changes in practice, without consequences for teachers or students, enabling authentic collaboration and time for values to be explored, reflected upon, and discussed are essential processes.

Having an expectation that change may be a difficult process makes it easier to embrace challenges and see them as a source of creativity and motivation to continue (Gunter, 1997). Transformations in practice rely on the emotional engagement of teachers and students as well as their intellectual engagement, so striving for changed conceptions of teaching and learning rather than superficial alterations in practice alone is a worthy goal (Trigwell and Prosser, 1997).

Moving to Student-Centered Pedagogy in Mixed-Mode Education

In a face-to-face situation, alerting students to a new teaching practice may be as simple as indicating the approach that the teacher intends to take and why this approach is appropriate to the learning objectives intended for students. The articulation of beliefs and the development of a shared view of practice and its implications can be managed relatively easily through group discussion. Concurrently, students may be encouraged to discuss the purpose and relevance of the new teaching strategies so that they become increasingly familiar with and accepting of the introduced practice. Nonverbal communication, an integral part of this process, is possible when teachers and students are colocated. It reduces the potential for ambiguity and heightens the psychological engagement of both parties, potentially leading to a positive outcome (Kock, 2002).

In the distance mode, motivating students and developing a shared view is made more difficult by the relative isolation of individuals who comprise the group (Gravoso, 2002). Traditionally, communicating the teaching strategy and supporting students to reach acceptance of it has relied on one-way communication, usually written, with some supplementation from audio tapes. Where broadband technologies can be utilized, the potential for a mix of indirect and virtual communication provides an opportunity to enhance the process through the use of rich media (Kock, 2001). The choice of media can include opportunities for interactive online discussion, videoconferencing, audio conferencing, and group interaction through e-mail. Virtual learning circles and mentoring are strategies that can utilize rich media to provide support for isolated students. The lecturer's thoughtfulness about changing the teaching approach will be contingent upon consideration of available communication media, their relative costs, and their suitability for the context and audience (Green, 2002; Jones and Richardson, 2002; Koenders, 2002). A range of theoretical and practical perspectives drawn from current debate surrounding the appropriate use of ICTs informs the remainder of the following discussion (Collis and Moonen, 2001; Roberts et al., 2002).

TEACHING FOR STUDENT ENGAGEMENT IN SCIENCE

For many students, one of the appeals of studying the sciences can be the practical, hands-on nature of many learning experiences. For students studying at a distance, the lack of opportunities for such experiences, without the significant expense of attending intensive, on-campus classes, can be a disadvantage (Rowntree, 1994). One means to increase practical interaction could be through the use of videoconferencing to enhance distance students' experiences by engaging them with their peers in interactive tutorials as a regular part of their learning (Blake and Taji, 1997; Gratton, 1998). These virtual tutorials or activities may involve both distance and on-campus learners.

The introduction of broadband technology will improve the potential for the videoconferencing medium because it increases richness by allowing video and audio transmission in multiple streams of images to be conveyed to multiple sites simultaneously (Fillion, Limayem, and Bouchard, 1999; Pitcher, Davidson, and Goldfinch, 2000; Motamedi, 2001). Recently, a report of research into the influence of students' perceptions showed that task orientation, equity, and innovative teaching methods are aspects of the learning environment valued among tertiary science students (Nair and Fisher, 2001). If well implemented, the introduction of videoconferencing has the potential to engage students meaningfully in learning experiences and tasks and to demonstrate innovative pedagogy and equity. However, such a transition should be approached with caution, if evidence from secondary science classrooms is an indicator (Stolarchuk and Fisher, 2001). A framework for making decisions about the appropriate uses of technology within particular pedagogical contexts should be considered (Laurillard, 2002). In considering the usefulness of broadband Internet-based videoconferencing, the following discussion draws upon wisdom from emerging literature to focus on the use of sound pedagogy and interactive student-centered learning (Herrington and Bunker, 2002; Herrington and Oliver, 2002).

EXPLORING THE USEFULNESS
OF BROADBAND VIDEOCONFERENCING

The choice of media for particular aspects of a unit or course should depend upon the purpose for teaching and the needs of the learners (Laurillard, 2002). Consideration should be given to the need for, or usefulness of, various media within a discipline, the optimum frequency of use, and the fit of the media to the circumstances of the learners (Gilman and Turner, 2001). In the following discussion, the term *broadband Internet-based videoconferencing* is used as the descriptor for a configuration of small- and medium-sized systems capable of allowing synchronous, interactive visual communication between individuals and both small and large groups. It does not include synchronous Web conferencing using desktop systems but may be applied to individuals using desktop systems linked to a larger room where several students are gathered.

How Will This Be Useful in the Sciences?

Apart from successfully demonstrating such practices as tissue culture (Blake and Taji, 1997), videoconferencing has been used at the University of New England to engage remote school students learning about the physiology and life cycle of fish, to demonstrate aspects of wool science, for tutorials, postgraduate supervision, and vivas. These useful experiences have, however, generally been teacher-directed rather than student-centered because the medium lends itself particularly well to the one-to-many application (Laurillard, 2002). Very recent advances in the technology, which have made transmission equipment portable and wireless, now make it possible to utilize the richness of broadband Internet-based videoconferencing much more effectively (Kingham, 2002). It could be used as a means of including remote students in routine group or practical work, as well as on-campus activities and field trips. For example, a student on a life-support system many thousands of kilometers from the University of New England campus participated in a two-day residential school where students were actively engaged in problem-based learning in groups. This student interacted constantly with peers and effectively led one group.

The ability of end users to recognize body language as well as intonation of speech means that the medium has the potential to

- enable access to the multiple cues of communication that are present in natural language;
- approximate regular face-to-face situations for remote students usually unable to attend;
- enable opportunities for immediate feedback; and
- personalize learning for remote students (Gilman and Turner, 2001).

Added to this, the establishment of high-speed broadband networks makes use of broadband Internet-based videoconferencing cost-effective when compared to its landline-based predecessor (Kingham, 2002). Broadband technology allows three to eight times the quality in picture and sound, so that these approximate the images seen on television screens. Dual transmission of live action enhanced with simultaneous transmission of video, PowerPoint slides, and audio tracks is easily achieved. In the previous example, the remote student was able to observe his or her fellow students in the lecture theater, see the Internet sites that the lecturer was demonstrating, and respond to questions in real time (Scanlon, 2002).

This capability opens up possibilities previously unavailable to students by mixing broadband Internet-based videoconferencing and traditional teaching techniques. When investigating these techniques, teachers could aim to

- improve students' access to other students, thereby reducing the isolation of remote learners by facilitating the development of support networks;
- enhance the experiences of full-time students by providing them with opportunities to interact with students working professionally in the field;
- empower students by increasing the flexibility of learning situations so that students could choose to participate in real-time, archived, or face-to-face activities as their needs and finances allowed;

- engage remote students more fully, intellectually and emotionally, in their learning by combining videoconferencing with traditional learning activities and information communication technologies (Koenders, 2002); and
- become more inclusive for students with disabilities or limiting geographical/familial circumstances.

What Types of Active Learning Experiences May Be Possible?

Taken from recent literature concerning the use of ICTs generally, and in the sciences in particular, the possibility of utilizing broadband Internet-based videoconferencing for role-plays, interactive group work, simulation games, practical demonstrations, and more traditional activities, such as guest lectures and tutorials, seems probable (Blake and Taji, 1997; Gilman and Turner, 2001; Benbunan-Fich and Stelzer, 2002). Several elements related to the immediacy of the medium, such as the following, could enhance learning activities in the sciences:

1. With the aid of a large room system in the laboratory and smaller systems in tutorial spaces, the lecturer conducting a practical class could expect remote students to participate through questioning and discussion.
2. For problem-based scenarios, students could interact with one another and the lecturer; make presentations to distance education groups simultaneously; and utilize computer-based simulations, videotape of practical or field work, and even demonstrate techniques to one another. Sessions could be videotaped so that students unable to participate synchronously could access archived sessions via video streaming.
3. Similarly, remote students, acting as members of a virtual learning circle that includes on-campus members, could view and direct complex experiments or walk with students filming a field trip. Alternatively, remote students could view previous years' trips and participate in the planning of current trips so that activities of interest to their learning needs are available in virtual rather than real time.

Most important, person-to-person feedback and discussion will be possible at minimum cost to many more students than is currently the case, once broadband Internet-based videoconferencing networks are established across Australia.

Effective use of media can support learning by meeting the needs of the diverse body of students enrolling in higher education and open up possibilities for clarification, negotiation, collaborative feedback, and thoughtful evaluation of teaching and learning (Laurillard, 2002).

DEVELOPING A CONCEPTUAL FRAMEWORK TO PLAN POTENTIAL STUDENT ENGAGEMENT

If we accept that moving from a teacher-centered to a student-centered approach requires thoughtfulness, reflection, and planning, then it is probably wise to consider the use of a planning framework. Figure 11.1 presents an example of a conceptual framework that could be devised by a teacher deciding which types of pedagogical interactions might appropriately be used in particular contexts. In addition, a more extensive analysis of the time, place, synchronicity, and pedagogy should be used to ensure appropriate alignment between the needs of the learner, the learning context, the costs of available technology, and the purpose for teaching and learning (Collis and Moonen, 2001; Benbunan-Fich and Stelzer, 2002; Laurillard, 2002).

Figure 11.1 presents a series of possible interactions that could be facilitated by broadband Internet-based videoconferencing and attempts to fit them along a continuum of increasing student-centeredness. The descriptors used in each section of the matrix give examples of teaching and learning strategies that might be appropriate for the type of interaction and the degree of student-centeredness appropriate to a desired teaching context or purpose. The range of examples is intended to indicate the variety of ways in which the videoconferencing medium might enhance learning. It illustrates strategies from the more traditional teacher-directed lecture to dominantly student-directed activities.

The richness of the medium makes it possible for students to be involved in experiences ranging from dialogue to virtual participation

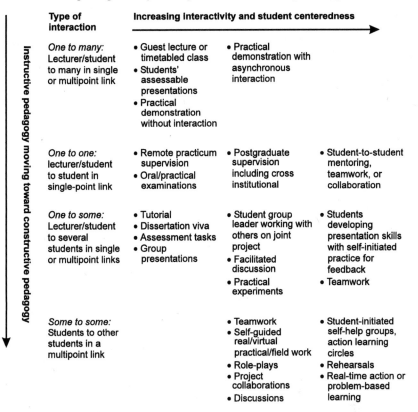

FIGURE 11.1. An Example of a Conceptual Framework for Planning Learner Engagement Using Videoconferencing (Single or dual video feed is possible for all interactions.)

in practical work. Choosing the experience and the level of interactivity will reflect the lecturer's approach to utilizing the new medium and teaching techniques. It is really a case of thoughtful analysis of learning needs, where the lecturer plans the level of student engagement according to the limitations of the task, the suitability of the medium, and its appropriateness to the content and the learning environment. Each form of experience can have differing levels of interactivity and student freedom, according to the ways in which the lecturer plans the students' engagement with learning materials and experiences (Laurillard, 2002).

CONCLUSION

This chapter has explored some of the challenges and issues arising out of the need to change tertiary teaching toward student-centered approaches. It has examined pedagogical and philosophical influences, issues associated with managing such a significant change in practice, and the potential of broadband-Internet-based videoconferencing as a means of enhancing mixed-mode delivery. Based on extant research, there is great potential for broadband Internet-based videoconferencing to add richness to learning environments in tertiary science.

REFERENCES

Adelman, N.E. and Walking Eagle, K.P. (1997). Teachers, time, and school reform. In A. Hargreaves (Ed.), *ASCD year book—Rethinking educational change with heart and mind* (pp. 92-110). Alexandria, VA: Association for Supervision and Curriculum Development.

Andresen, L.W. (2000). A useable, trans-disciplinary conception of scholarship. *Higher Education Research and Development,* 19(2): 137-153.

Benbunan-Fich, R. and Stelzer, L. (2002). Computer-supported learning of information systems: Matching pedagogy with technology. In E. Cohen (Ed.), *Challenges of information technology education in the 21st century* (pp. 85-99). Hershey, PA: Idea Group Publishing.

Biggs, J. (1999). *Teaching for quality learning at university: What the student does.* Buckingham, United Kingdom: Society for Research into Higher Education and Open University Press.

Biggs, J.B. and Collis, K.F. (1982). *Evaluating the quality of learning.* Sydney: Academic Press.

Blake, A. and Taji, A. (1997). Teaching plant tissue culture by videolink. In A. Taji and R. Williams (Eds.), *Tissue culture toward the next century* (pp. 173-184). Armidale, Australia: University of New England.

Boyer, E.L. (1990). *Scholarship reconsidered: Priorities of the professoriate.* Princeton, NJ: The Carnegie Foundation for the Advancement of Teaching.

Buckley, D.P. (2002). In pursuit of the learning paradigm: Coupling faculty transformation and institutional change. *EDUCAUSE,* (January/ February): 29-38.

Carr, W. and Kemmis, S. (1986). *Becoming critical: Education, knowledge, and action research.* London: The Falmer Press.

Collis, B. and Moonen, J. (2001). *Flexible learning in a digital world.* London: Kogan Page.

Cooksey, R. (2000). *Systems thinking and group problem-solving/decision making tools and techniques.* Armidale, Australia: University of New England.

Fillion, G., Limayem, M., and Bouchard, L. (1999). Videoconferencing in distance education: A study of student perceptions in the lecture context. *Innovations in Education and Training International,* 36(4): 302-310.

Foley, G. (Ed.) (2000). *Understanding adult education and training* (Second edition). Sydney: Allen and Unwin.

Fosnot, C.T. (Ed.) (1996). *Constructivism: Theory, perspectives, and practice.* New York: Teachers College Press.

Fullan, M. (1991). *The NEW meaning of educational change* (Second edition). London: Cassell Educational Ltd.

Fullan, M. (1997a). The complexity of the change process. In M. Fullan (Ed.), *The challenge of school change: A collection of articles* (pp. 27-46). Highett, Australia: Hawker Brownlow Education.

Fullan, M. (1997b). Leadership for change. In M. Fullan (Ed.), *The challenge of school change: A collection of articles* (pp. 97-114). Highett, Australia: Hawker Brownlow Education.

Fullan, M. (1998). Leadership for the 21st century: Breaking the bonds of dependency. *Educational Leadership,* 55(7): 6-10.

Gibbs, G. (1995). Changing lecturers: Conceptions of teaching and learning through action research. In A. Brew (Ed.), *Directions in staff development* (pp. 21-35). Buckingham, United Kingdom: Society for Research into Higher Education and Open University Press.

Gilman, S.C. and Turner, J.W. (2001). Media richness and social information processing: Rationale for multifocal continuing medical education activities. *Journal of Continuing Education in the Health Professions,* 21(3): 134-138.

Gratton, L. (1998). *Video conferencing.* Armidale, Australia: The Teaching and Learning Centre, University of New England.

Gravoso, R. (2002). Meeting the need for relevance and quality learning: A case in using the constructivist learning approach. In A. Goody, J. Herrington, and M. Northcote (Eds.), *Research and development in higher education: Quality conversations* (pp. 290-297). Perth, Australia. Proceedings, July 7-10. Higher Education Research and Development Society of Australasia Inc. (HERDSA).

Green, J. (2002). Exemplars of on-line peer support—Are we looking in the right places? In A. Goody, J. Herrington, and M. Northcote (Eds.), *Research and development in higher education: Quality conversations* (pp. 298-304). Perth, Australia. Proceedings, July 7-10. Higher Education Research and Development Society of Australasia Inc. (HERDSA).

Grundy, S. (1987). *Curriculum: Product or praxis.* Melbourne: The Falmer Press.

Grundy, S. (1992). Beyond guaranteed outcomes: Creating a fiscourse of praxis. *Australian Journal of Education,* 36(2): 157-169.

Gunter, H. (1997). Chaotic reflexivity. In M. Fullan (Ed.), *The challenge of school change: A collection of articles* (pp. 71-96). Highett, Australia: Hawker Brownlow Education.

Hannah, W. (1979). John Dewey: Education for intelligently directed action. In J.V. D'Cruz and W. Hannah (Eds.), *Perceptions of excellence* (pp. 115-144). Melbourne: The Polding Press.

Hargreaves, A. (1996). Revisiting voice. *Educational Researcher,* 25(1): 12-19.

Hargreaves, A. (1997a). Cultures of teaching and educational change. In M. Fullan (Ed.), *The challenge of school change: A collection of articles* (pp. 47-68). Highett, Australia: Hawker Brownlow Education.

Hargreaves, A. (1997b). The four ages of professionalism and professional learning. *Unicorn,* 23(2): 86-114.

Hargreaves, A. (1997c). Rethinking educational change: Going deeper and wider in the quest for success. In A. Hargreaves (Ed.), *ASCD year book—Rethinking educational change with heart and mind* (pp. 1-26). Alexandria, VA: Association for Supervision and Curriculum Development.

Hargreaves, A. (1998). Emotions of teaching and educational change. In A. Hargreaves, A. Lieberman, M. Fullan, and D. Hopkins (Eds.), *International handbook of educational change,* Volume 1 (pp. 558-575). Dordrecht, the Netherlands: Kluwer Academic Publishers.

Hargreaves, A., Lieberman, A., Fullan, M., and Hopkins, D. (Eds.) (1998). *International handbook of educational change,* Volume 1. Dordrecht, the Netherlands: Kluwer Academic Publishers.

Herrington, A. and Bunker, A. (2002). Quality teaching online: Putting pedagogy first. In A. Goody, J. Herrington, and M. Northcote (Eds.), *Research and development in higher education: Quality conversations* (pp. 305-312). Perth, Australia. Proceedings, July 7-10. Higher Education Research and Development Society of Australasia Inc. (HERDSA).

Herrington, J. and Oliver, R. (2002). Designing for reflection in online courses. In A. Goody, J. Herrington, and M. Northcote (Eds.), *Research and development in higher education: Quality conversations* (pp. 313-319). Perth, Australia. Proceedings, July 7-10. Higher Education Research and Development Society of Australasia Inc. (HERDSA).

Jones, S. and Richardson, J. (2002). Designing an IT-augmented student-centered learning environment. In A. Goody, J. Herrington, and M. Northcote (Eds.), *Research and development in higher education: Quality conversations* (pp. 376-383). Perth, Australia. Proceedings, July 7-10. Higher Education Research and Development Society of Australasia Inc. (HERDSA).

Kingham, S. (2002). CSIRO case study: Facilitating collaboration using video over IP conferencing. Questnet. Available at <http://www.aarnet.edu.au/engineering/wgs/video/presentations/>.

Kirkpatrick, D. and Jakupec, V. (1999). Becoming flexible: What does it mean? In A. Tait and R. Mills (Eds.), *The convergence of distance and conventional education: Patterns of flexibility for the individual learner* (pp. 51-70). London and New York: Routledge.

Kock, N. (2001). Compensatory adaptation to a lean medium: An action research investigation of electronic communication in process improvement groups. *IEEE Transactions on Professional Communication,* 44(4): 267-285.

Kock, N. (2002). *Media richness or media naturalness? The evolution of our biological communication apparatus and its influence on our behaviour toward e-communication tools* [Research report ERC-2002-2]. Philadelphia: Temple University.

Koenders, A. (2002). Creating opportunities from challenges in on-line introductory biology. In A. Goody, J. Herrington, and M. Northcote (Eds.), *Research and development in higher education: Quality conversations* (pp. 393-400). Perth, Australia. Proceedings, July 7-10. Higher Education Research and Development Society of Australasia Inc. (HERDSA).

Laurillard, D. (2002). *Rethinking university teaching: A framework for the effective use of educational technology.* London: Routledge.

Louis, K.S., Toole, J., and Hargreaves, A. (1999). Rethinking school improvement. In J. Murphy and K.S. Louis (Eds.), *Handbook of research on educational administration* (pp. 251-276). San Francisco: Jossey-Bass.

Mezirow, J. (Ed.) (1990). *Fostering critical reflection in adulthood: A guide to transformative and emancipatory learning.* San Francisco: Jossey-Bass.

Mezirow, J. (1991). *Transformative dimensions of adult learning.* San Francisco: Jossey-Bass.

Mitchell, G.C. and Hope, B.G. (2002). Teaching or technology: Who's driving the bandwagon? In E. Cohen (Ed.), *Challenges of information technology education in the 21st century* (pp. 125-144). Hershey, PA: Idea Group Publishing.

Motamedi, V. (2001). A critical look at the use of videoconferencing in United States distance education. *Education,* 122(2): 386-394.

Nair, C.S. and Fisher, D.L. (2001). Learning environments and student attitudes to science at the senior and tertiary levels. *Issues in Educational Research,* 11(2): 12-30.

Panizzon, D., Pegg, J., and Mulquiney, C. (1999). Changing perspectives in tertiary teaching: A collaborative approach between science and education faculties. Australian Association for Research in Education. Available online at <http://www.aare.edu.au/99pap/pan99576.htm>.

Pitcher, N., Davidson, K., and Goldfinch, J. (2000). Videoconferencing in higher education. *Innovations in Education and Training International,* 37(3): 199-206.

Quality Assurance Agency (2000). Guidelines on the quality assurance of distance learning. Quality Assurance Agency for Higher Education (UK). Available online at <http://www.qaa.ac.uk/Public/dlg/contents.htm>.

Ramsden, P. (1992). *Learning to teach in higher education.* London: Routledge.

Roberts, J., Brindley, J., Mugridge, I., and Howard, J. (2002). Faculty and staff development in higher education: The key to using ICT appropriately? The Observatory on Borderless Education. Available online at <http://www.obhe.ac.uk/products/reports/>.

Rowntree, D. (1994). *Preparing materials for open, distance, and flexible learning: An action guide for teachers and trainers.* London: Kogan Page.

Salmon, D. (1998). *Using email and Listservs in teaching.* Armidale, Australia: The Teaching and Learning Centre, University of New England.

Scanlon, J. (2002). Video link takes disabled student to UNE. University of New England, Australia. September 30. Press Release number 120/2002 url: <http://www.une.edu.au/news/releases2002/September/120-02.html>.

Sergiovanni, T. (1998). Market and community as strategies for change. In A. Hargreaves, A. Lieberman, M. Fullan, and D. Hopkins (Eds.), *International handbook of educational change,* Volume 1 (pp. 576-595). Dordrecht, the Netherlands: Kluwer Academic Publishers.

Smyth, R. (2002). Knowledge, interest, and the management of educational change. Unpublished doctoral dissertation. University of New England, Armidale, Australia.

Smyth, R. (2003). Concepts of change: Enhancing the practice of academic staff development in higher education. *International Journal of Academic Development,* 8(1-2): 51-60.

Stolarchuk, E. and Fisher, D. (2001). First years of laptops in science classrooms result in more learning about computers than science. *Issues in Educational Research,* 11(1): 25-40.

Trigwell, K. and Prosser, M. (1997). Toward an understanding of individual acts of teaching and learning. *Higher Education Research and Development,* 16(2): 241-252.

Trigwell, K., Martin, E., Benjamin, J., and Prosser, M. (2000). Scholarship of teaching: A model. *Higher Education Research and Development,* 19(2): 156-168.

Trigwell, K. and Prosser, M. (1999). *Understanding learning, and teaching: The experience in higher education.* Buckingham, United Kingdom: Society for Research into Higher Education and Open University Press.

Twigg, C. (2002). Improving quality and reducing costs: Designs for effective learning using information technology. The Observatory on Borderless Higher Education. Available online at <http://www.obhe.ac.uk/products/reports/>.

von Glasersfeld, E. (1995). *Radical constructivism: A way of knowing and learning.* London: The Falmer Press.

Webb, G. (1996). *Understanding staff development.* Buckingham, United Kingdom: Society for Research into Higher Education and Open University Press.

Wylie, A. (1998). *Using voicemail for teaching and learning.* Armidale, Australia: The Teaching and Learning Centre, University of New England.

Index

Page numbers followed by the letter "b" indicate boxed material; those followed by the letter "i" indicate illustrations; and those followed by the letter "t" indicate tables.

Students *(continued)*
  dPBL exercises, 95
  Faculty Mentor Program, 131-133,
    133b, 134b, 136b
  first-year challenges, 123-124, 178
  first-year experience, 157
  first-year learning supports, 125,
    126t, 127-130
  in higher education, 4
  and laboratory work, 43
  learning process, 119t
  mentoring, 125, 129-130
  MetAHEAD evaluation, 78-79, 80b
  misconceptions of, 110
  and pedagogical change, 205
  retention, learning support, 5
  on SAMs' usefulness, 172, 173t
  and science research, 1
  statistics experience, 146b
  in Warwick ecology course, 13
Studios, laboratory learning, 46-47
Summary evaluation, dPBL exercises,
  94, 95
Summative assessment
  definition of, 179t
  and feed back, 160, 161
Survey design, learning opportunities, 6
Survey design and biostatistics
  biostatistics experience, 145
  statistics, 143
Surveys
  Faculty Mentor Program, 132, 133b,
    134b
  Griffith University study, 49-50,
    50-51, 55
  SAMs evaluation, 171-172, 172t,
    173t
Sustainable development code (UK),
  16-17

Taxonomy of Educational Objectives, 168
Taylor, Charlotte, 2
Teaching
  casual (adjunct) faculty, 159
  changed paradigm, 4

Teaching *(continued)*
  dPBL, 86
  holistic examination, 199
  and ICT skills, 200-201
  as interventionist activity, 2
  PBL, 85
  pedagogical change, 201-203
  problem-solving strategies, 117-119,
    119t
  reinventing the university, 157
  and science research, 1
Teaching and Learning Program
  (TLTP), software
  development, 27-28
Teaching materials, 16-17
Teaching strategies
  and ICT, 205-206, 208-209
  questioning, 114, 114t
  self-knowledge support, 111, 111t
  survey design/biostatistics course, 6,
    143, 149-152
  student-centered strategies, 210-211,
    211t
Teamwork
  Land course, 90
  PBL, 85
Technology
  benefits of, 6-7
  and constructivist pedagogy, 4
  reflection enhancements, 71-72
Tertiary literacy
  definition of, 123
  first-year students, 124
  student support, 4
Theory and practice, laboratory
  learning, 47-48
Think aloud
  problem-solving modeling, 110
  problem-solving skills, 105
  problem-solving strategies, 7-8
Time
  alternative conceptions, 148-149
  educational challenges, 2
  Warwick ecology course, 12
Time-management issues, first-year
  students, 132, 133

Order a copy of this book with this form or online at:
http://www.haworthpress.com/store/product.asp?sku=5294

TEACHING IN THE SCIENCES
Learner-Centered Approaches

_____in hardbound at $39.95 (ISBN: 1-56022-263-8)

_____in softbound at $29.95 (ISBN: 1-56022-264-6)

Or order online and use special offer code HEC25 in the shopping cart.

COST OF BOOKS_____

☐ **BILL ME LATER:** (Bill-me option is good on US/Canada/Mexico orders only; not good to jobbers, wholesalers, or subscription agencies.)

☐ Check here if billing address is different from shipping address and attach purchase order and billing address information.

POSTAGE & HANDLING_____
*(US: $4.00 for first book & $1.50
for each additional book)*
*(Outside US: $5.00 for first book
& $2.00 for each additional book)*

Signature_____

SUBTOTAL_____

☐ **PAYMENT ENCLOSED: $**_____

IN CANADA: ADD 7% GST_____

☐ **PLEASE CHARGE TO MY CREDIT CARD.**

STATE TAX_____
*(NJ, NY, OH, MN, CA, IL, IN, & SD residents,
add appropriate local sales tax)*

☐ Visa ☐ MasterCard ☐ AmEx ☐ Discover
☐ Diner's Club ☐ Eurocard ☐ JCB

FINAL TOTAL_____

Account #_____

*(If paying in Canadian funds,
convert using the current
exchange rate, UNESCO
coupons welcome)*

Exp. Date_____

Signature_____

Prices in US dollars and subject to change without notice.

NAME_____

INSTITUTION_____

ADDRESS_____

CITY_____

STATE/ZIP_____

COUNTRY_____ COUNTY (NY residents only)_____

TEL_____ FAX_____

E-MAIL_____

May we use your e-mail address for confirmations and other types of information? ☐ Yes ☐ No
We appreciate receiving your e-mail address and fax number. Haworth would like to e-mail or fax special
discount offers to you, as a preferred customer. **We will never share, rent, or exchange your e-mail address
or fax number.** We regard such actions as an invasion of your privacy.

Order From Your Local Bookstore or Directly From
The Haworth Press, Inc.
10 Alice Street, Binghamton, New York 13904-1580 • USA
TELEPHONE: 1-800-HAWORTH (1-800-429-6784) / Outside US/Canada: (607) 722-5857
FAX: 1-800-895-0582 / Outside US/Canada: (607) 771-0012
E-mailto: orders@haworthpress.com

For orders outside US and Canada, you may wish to order through your local
sales representative, distributor, or bookseller.
For information, see http://haworthpress.com/distributors

(Discounts are available for individual orders in US and Canada only, not booksellers/distributors.)
PLEASE PHOTOCOPY THIS FORM FOR YOUR PERSONAL USE.
http://www.HaworthPress.com   BOF04